AF708484

MONTAGNE

...NONYME

...deries de Zinc

...-MONTAGNE.

...'Action A

...nominale de Huit cents francs)

...51 AU PORTEUR.

BELGIQUE 6 Fr. BELGIE

... Mai 1927.

L'Administrateur-Directeur Général,

CARBONNEAU. SC.

...ONT (Cl.). — 6.4.1927.

Minerals with Belgian Roots

from hopeite (1824) to tazieffite (2009)

Minerals with Belgian Roots
from hopeite (1824) to tazieffite (2009)

Eddy Van Der Meersche
Paul De Paepe
Georges Stoops

UNIVERSITEIT GENT
ACADEMIA PRESS

Editor:

Alexander Jonckheere, Kristel Wautier and Danny Segers
Museum for the History of Sciences
Krijgslaan 281 S30
B-9000 Gent
Belgium
T (+32) (0)9 264 49 30 F (+32) (0)9 264 49 73
sciencemuseum@ugent.be www.sciencemuseum.UGent.be

Publisher:

© Academia Press
Eekhout 2
B-9000 Gent
T (+32) (0)9 233 80 88 F (+32) (0)9 233 14 09
info@academiapress.be www.academiapress.be

The publications of Academia Press are distributed by:

Belgium:
J. Story-Scientia Scientific Booksellers
Sint-Kwintensberg 87
9000 Gent
T (+32) (0)9/225 57 57 F (+32) (0)9/233 14 09
info@story.be www.story.be

The Netherlands:
Ef & Ef
Eind 36
NL-6017 BH Thorn
T (+31) (0)475 561501 F (+31) (0)475 561660

Rest of the world:
UPNE, Lebanon, New Hampshire, USA
(www.upne.com)

Design and Layout:
Alexander Jonckheere
Eddy Van Der Meersche

Eddy Van Der Meersche (e.vandermeersche@skynet.be)
Paul De Paepe (paul.depaepe@ugent.be)
Georges Stoops (stoops.georges@skynet.be)
Minerals with Belgian roots. From hopeite (1824) to tazieffite (2009)
Gent, Academia Press, 2010, 232 pp.

ISBN 978 90 382 1531 0
D/2010/4804/1
U 1377

The authors retain all responsability for the content of this publication.

No part of this publication may be reproduced in print, by photocopy, microfilm or any other means, without the prior written permission of the publisher.

EDITORIAL NOTE

"Minerals with Belgian Roots" - The birth of an idea

The Museum for the History of Sciences at Ghent University houses an extensive collection of scientific instruments that have been used at our university in teaching and research. Date wise, the collection covers over 200 years, from the establishment of the university in 1817 up till the end of the 20th century. This Museum is found on the campus of the Science Faculty of Ghent University. It shows a permanent exposition of instruments relating to different disciplines of science such as physics, chemistry, biology, geology, geography, ... On a regular basis, the Museum also organizes temporary exhibitions on various topics and related to different scientific disciplines.

In 2006, I lunched with some colleagues from the science campus in a nearby restaurant. Georges Stoops and Paul De Paepe, both retired professors from the Geology and Soil Science, were amongst those colleagues. Both were still working on a voluntary basis at their former laboratory on a voluntary basis. As such, they were still involved in the ongoing scientific research.

During that lunch, they inquired about the next temporary exhibitions in the Museum. I briefly outlined the plans for the near future and suggested that this could include doing something about geology as well. Although there was no immediate response, this may have inspired them.
Indeed, when we met again some time later, Paul De Paepe and Georges Stoops proposed to work around minerals related to Belgium. They pointed out, that in the past some renowned authorities from the geology department of our university studied minerals. They were also of the opinion that there must be minerals with names reflecting Belgian personalities or Belgian localities.

This marked the birth of a new idea! From then onwards, the museum incorporated such a temporary exhibition into its planning. We also started looking for the necessary funds. In the meanwhile, the geologists started their research. They also got in touch with Eddy Van Der Meersche and asked him to join their team. Like Paul De Paepe and Georges Stoops, Eddy was also formed at our university as a geologist. Yet, he had chosen a career as a secondary school teacher and was by now retired. During the years, Eddy had acquired a lifelong experience in mineral photography and is considered an excellent photographer amongst mineralogists.

Together, they started planning and worked out a general outline. Gradually, the idea of a book accompanying the exhibition grew. After a long period of hard work, their ideas finally took shape. When it was decided on March 4th, 2010 for the official opening of the exposition, this also implied the final deadline for the production of this book. As time went by we often wondered whether we would ever be able to meet this deadline. In the end, we did and the result is this magnificent book. We hope you enjoy reading it!

Many thanks to all the authors for their splendid work and their support in organizing the exhibition.

Ghent, January 15th 2010

Prof. Danny Segers
Director Museum for the History of Sciences - Ghent University

PREFACE

Minerals occurring in Belgium have amply and comprehensively been documented in books by Buttgenbach (1947), Mélon *et al.* (1976) and Hatert *et al.* (2002), see references in the introductory chapter of the current book. These three review books show the evolution of mineralogical science and knowledge in Belgium through the last 60 or so years. As most books on mineralogy, however, also the Belgian books lack extensive information on the etymology of mineral names. And moreover, these books do not list minerals occurring ouside of Belgium, but named after Belgian individuals.

The voting procedure on proposals for a new mineral as imposed by the Commission on New Minerals, Nomenclature and Classification (CNMNC) of the International Mineralogical Association (IMA) consists of two parts: a vote on the descriptive data presented by the author(s) for the presumably new mineral, and a vote on the name proposed by the author(s) for this mineral, if indeed approved as being new to science. There is an enormous discrepancy in the importance of those two votes. Whereas the former vote, on the data, understandably requires many pages of detailed data to fully inform the CNMNC members, the latter vote, on the name, regrettably is cast on a single paragraph containing only the barest minimum of information on the reasons for the new name.

Although the CNMNC has a set of guidelines for mineral nomenclature, it could alas not prevent the establishment of mineral names which might be viewed as inconsiderate, such as bakhchisaraitsevite, pääkkönenite, uchacchacuaite and zhemchuzhnikovite. Fair is fair, many persons outside of Belgium (or even within Belgium?) might equally have some difficulties in pronouncing correctly the names vandendriesscheite or vanmeersscheite. But Johan Wolfgang von Goethe, not only a world-class writer but also a keen mineralogist, stated already in 1807 that giving names to minerals is not as easy as one thinks.

In the last few years a new kind of national mineralogy books has originated. These books are not only compilations of minerals found in a given country, they also give information on the etymology of the mineral names, in the case of named after persons illustrated with full biographies and photographs, and information on minerals named after individuals from this particular country. Such books have been published for the former Soviet Union, Canada, Greenland, Switzerland and Italy, they greatly increase the background knowledge of mineralogy, especially on the individuals connected with it.

Belgium could not be missing in this series of books. Because it is a relatively small country and because there have been (and still are) so many eminent and prolific mineralogists, in terms of density of new mineral species per surface unit, Belgium occupies world-wide a prominent fifth place (after Switzerland, Czech Republic, Italy and Germany) in such a ranking.

Minerals are not only scientific or aesthetical objects, they also are a part of the country's patrimony. The authors of this book, in providing so many extensive and new informations on these somewhat unappreciated national treasures, are to be congratulated for their important contribution to Belgium's mineralogical heritage.

Ernst A.J. Burke
Chairman (2003-2008) of the IMA-CNMNC.

CONTENTS

WHAT'S IN A NAME?

1. Minerals with Belgian roots

In most handbooks on systematic mineralogy very little, if any, information is given on the etymology of mineral names. Moreover, in the case of minerals named after persons information is mostly restricted to the name of that person, sometimes completed with the year of his birth and death, and his profession. Information on his education and career is generally difficult to find. It was the aim of the authors of this book to provide information on the names of minerals first discovered in Belgium, or abroad and named after Belgian individuals. Not only names of those minerals accepted by the International Mineralogical Association (IMA) are discussed, but also older, discredited names, as they are part of the national science history. More general information on mineral names can be found in Chester (1896), Mitchell (1979), Cesbron (1985), Blackburn & Dennen (1997) and de Fourestier (1999).

It is clear that this book has a historical bibliographical background, and is in no way meant as a mineralogical descriptive handbook. Only some basic data are given, and for more information on the minerals the reader is referred to recent handbooks, such as that authored by Anthony *et al*. (1990-2003), or the corresponding website http://pubsites.uws.edu.au/ima-cnmnc.

Belgian scientific institutes, such as universities, the Royal Belgian Institute of Natural Sciences in Brussels and the Royal Museum for Central Africa in Tervuren played an important role in descriptive mineralogy. Scientists of these institutions discovered, described and named many minerals, not only of Belgian origin, but also originating from foreign countries, especially in central Africa. Also the contributions of mining organizations, such as *Union Minière du Haut-Katanga*, should not be forgotten. It is evident that much more minerals were discovered and identified by Belgian researchers, but as they neither originated from Belgium, nor were named after Belgian personalities they are not discussed in this book (for instance becquerelite, curite, shabaite).

The minerals of Belgium were described in three comprehensive books written by Buttgenbach (1947), Mélon *et al.* (1976) and Hatert *et al.* (2002). These books are not only interesting as a guide to the occurrence and characteristics of minerals found in Belgium, but give together an idea of the historical evolution of our knowledge; their target audience comprises both professionals and amateurs.

2. Mineral names

Since the early days of mankind, rock- and mineral material has been used by our ancestors, as can be deduced from numerous excavated remains of stone tools, jewels and pigments. They will surely have had names for the different materials, although differentiation between rocks and minerals, as known today, is merely a recent phenomenon from a historic point of view.

Already in Egyptian and Babylonian texts names of minerals and rocks are quoted, especially of gemstones. From antiquity several treatises comprising information on materials reached us, such as Theophrastus' "On Stones" (c. 315 BC), Plinius the Elder's "Natural History" (c. 77 AD) and Vitruvius' treatises (first century BC). With regard to the Renaissance we especially have to mention the work of the German scholar and scientist Georg Bauer, better known as Georgius Agricola, who described many minerals in his book "De Re Metallica" (1556). The classification proposed by the Flemish humanist Anselmus Boetius De Boodt was also an important step in mineralogy (see: *boodtite*). The French traveller and pioneer of trade with India, Jean-Baptiste Tavernier, Baron of Aubonne (1605-1689), born in Paris as son of a Protestant family from Antwerp, contributed to mineralogy by his description of Indian gemstones.

With the rise of geology and mineralogy as a science in the 18th century, scientists made a distinction between minerals and rocks, the latter composed of many individual minerals of the same nature (e.g. in marble) or of different nature (e.g. in granite). Crystallographical properties and chemical composition were the first scientific criteria used to discriminate different mineral species. Introduction of more sophisticated methods in modern times, such as thermal analysis, polarizing microscopy and especially X-ray diffraction led to a multiplication of the number of new minerals discovered. Whereas in the old times minerals were known by traditional local names, given by miners (e.g. quartz, "mispickel", the French name for arsenopyrite, or "Grünbleierz", the German name for pyromorphite) or traders, the names of newly discovered minerals were more artificial, ending on the suffix -tite or -ite, from the Greek word λιθος, meaning stone. An example of a new name based on an old root is halite (NaCl) (compare with the numerous toponyms such as Hallein, Solbad Hall, Hallstad, all places where salt was mined).

For native elements the name of the element is used, which is different in every language, for instance: gold, goud (Dutch), or (French), Gold (German), oro (Spanish), or silver, zilver (Dutch), argent (French), Silber (German), plata (Spanish), or iron, ijzer (Dutch), fer (French), Eisen (German), hierro (Spanish).

In the beginning, minerals were mostly named after specific properties such as colour (e.g. rhodochrosite, for pinkish; azurite, for blue), density (e.g. barite, for heavy), shape (e.g. hemimorphite, which means half shape), etc. Later, it became common practice to use names of type localities (e.g. ardennite, ottrélite), personalities (e.g. goethite, after the German mineralogist and writer Johan Wolfgang von Goethe, or curite after the French physicist Pierre Curie), or chemical composition [e.g. calclacite for **cal**(cium) **c**(h)**l**(orine) and **ac**(etate) or nahcolite, for $NaHCO_3$].

The prefix meta- is used to indicate a mineral with a same chemical composition, but less structural water. Examples are schoepite and metaschoepite, vandendriesscheite and metavandendriesscheite. The prefix para- indicates generally a polymorph, for instance fransoletite (monoclinic) and parafransoletite (triclinic), hopeite (orthorhombic) and parahopeite (triclinic). In the case of minerals of rare earth elements, the most abundant of these elements is indicated between brackets as a suffix, for instance lepersonnite-(Gd), stavelotite-(La), françoisite-(Ce) and françoisite-(Nd). The same procedure is used to distinguish similar minerals with different composition such as ardennite-(V) and ardennite-(As).

As there was no international coordination, the same mineral could have a different name in Anglo-Saxon countries and for instance in French or German influenced areas. A rather striking example dates from the cold war. A fibrous magnesium silicate was named attapulgite in the USA (after the town Attapulgus, Georgia) and palygorskite in Russia (after the locality Palygorskaya, Ural Mountains), the latter name now being internationally accepted.

Not only were different names given to the same mineral species, but also minerals that only represented a variety of a mineral species (e.g. *bastonite*), or a member of a continuous series (e.g. the plagioclase series) got a name, with for the latter different limits of composition in different countries. In 1959, to put an end to this confusion, the International Mineralogical Association (IMA) decided to set up rules and procedures for naming new minerals, and to control and standardize all existing mineral names (so-called grandfathered or pre-IMA names). New names referring to an already known mineral, or mineral variety, were discredited. Examples are *dewalquite*, that was discredited because the same mineral was already named ardennite, and *bastonite* that is no longer considered as a separate mineral, but only as a variety of biotite. In this discussion, the chronology of the first publication plays an important role. Especially the rationalization of the nomenclature of mineral groups with more than two end members (e.g. pyroxenes, amphiboles, chlorites) took a lot of time and energy of special IMA-commissions.

To get a new mineral name approved, the mineralogist has to follow a rather long and strict procedure (Fleischer, 1970; Nickel & Grice, 1998). All data necessary to prove the discovery of a supposed new mineral species should be sent to the President of the "Commission on New Minerals and Mineral Names" of the International Mineralogical Association (IMA). The members of the Commission have to comment on the proposal and vote on the acceptation of the new name. For opaque minerals, the members of the "Ore Commission" of the IMA are consulted. Requested data comprise for instance chemical composition, optical properties, X-ray diffraction patterns and cell parameters.

Originally, names of internationally reputed, generally deceased scientists were given (e.g. braggite, curite, danalite, brewsterite, halloysite). Later, it became a trend in Belgium to honour a person of the same company or institute of the researcher, be it a senior geologist or mineralogist (e.g. cornetite, lepersonnite-(Gd), schoepite) or a senior administrator (e.g. cattierite, cousinite, marthozite, sengierite). The last decennia names of a colleague of the same research team (e.g. blatonite, deliensite, piretite, vochtenite) are often given. For names derived from persons, normally only the surname is used. A special case is mélonjosephite (see later), named after Joseph Mélon, where the first name was added to avoid confusion with melonite, named after the Californian locality Melones. Other examples are andrémeyerite and oswaldpeetersite.

Classification and terminology in mineralogy is quite different from that in life sciences, where the system of Carolus Linnaeus is applied, and plants and animals are grouped in orders, families, etc., each getting a scientific name generally illustrating the genetic relationship. Such a name always consists of two words, the first referring to the genus, the second the species. In mineralogy no such standardized names exist, and mineral names do not indicate relationships between minerals. A notable exception is the old medieval German subdivision of sulphides in "Kiese" (light, with metallic lustre), "Glanze" (dark, with metallic lustre), "Blenden" (coloured, without metallic lustre) and "Fahle", where the mineral name (e.g. Kupferkies, Bleiglanz, Zinkblende) immediately pointed to a subgroup level of the sulphides.

Varieties of minerals often get popular names, for instance different varieties of quartz such as amethyst (purple), agate (banded fine microcrystalline), morion (black), etc. They are not subject to strict rules. This is also the case for names of gemstones that often bear witness of a strong fantasy! Examples are the names given to different varieties of corundum, such as sapphire (blue), oriental emerald (green), ruby (red).

No generally accepted hierarchical classification of minerals exists. In most manuals minerals are traditionally grouped according to their chemical composition: native elements, sulphides, halogenides, carbonates, sulphates, etc. and finally the silicates. For the latter, subdivisions are based on the structure of the lattices. In other books the classification of all minerals is based on their lattice types.

Not all names ending on -ite point to minerals. Also many rocks were named according to their type locality (e.g. harzburgite, limburgite, monchiquite), and even artificial materials got such names. To give a Belgian example, beringite is the name for a cyclonic ash, used for soil amelioration, named after the municipality of Beringen in the Belgian province of Limburg. Another example is astridite, a chromojadeite (rock) named after the former Belgian Queen Astrid.

References

Anthony, J.W., Bideaux, R.A., Bladh, K. W. & Nichols M. C. (1990 – 2003) Handbook of Mineralogy (5 volumes). Tuscon, Arizona: Mineral Data Publishing.

Blackburn, W.H. & Dennen, W.H. (1997) Encyclopedia of Mineral Names. 360 p. *The Canadian Mineralogist*, Special Publication 1.

Buttgenbach H. (1947) Les Minéraux de Belgique et du Congo Belge. 573 p. Paris: Ed. Dunod.

Cesbron, F. (1985) World Directory of Mineralogists. 361 p. International Mineralogical Association.

Chester, A.H. (1896) A Dictionary of the Names of Minerals including the History and Etymology. London: Chapman & Hall, Ltd.

Clark, A.M. (1993) Hey's Mineral Index. Third Edition. 852 p. London: Chapman & Hall.

de Fourestier, J. (1999) Glossary of Mineral Synonyms. 434 p. *The Canadian Mineralogist*, Special Publication 2.

Deliens, M. (1988) Nouvelles tendances de la nomenclature minéralogique. *Bulletin des Séances de l' Académie royale des Sciences d'Outre-Mer,* **34,** 529-535.

Fleischer, M. (1970) Procedures of the International Mineralogical Association, Commission on New Minerals and Mineral Names. *The American Mineralogist,* **55,** 1016-1017.

Hatert, F., Deliens, M., Fransolet, A.-M. & Van Der Meersche, E. (2002) Les Minéraux de Belgique, 2e édition. 304 p. Bruxelles: Institut royal des Sciences naturelles de Belgique.

Mélon, J., Bourguignon, P. & Fransolet, A.-M. (1976) Les Minéraux de Belgique. 283 p. Dison: G. Lelotte.

Mitchell, R.S. (1979) Mineral Names: What do they Mean? 229 p. New York: Van Nostrand Reinhold Co.

Nickel, E.H. & Grice, J.D. (1998) The IMA Commission on New Minerals and Mineral Names: procedures and guidelines on mineralogical nomenclature. *The Canadian Mineralogist,* **36**, 1-16.

http://pubsites.uws.edu.au/ima-cnmnc.

KEY TO THE MINERAL DESCRIPTIONS

1. The main body of the present book consists of two parts. **Part I** deals with minerals discovered in Belgium: part of them having a name reminiscent of a Belgian individual, others with a name paying tribute to the place of discovery. **Part II** treats minerals discovered outside Belgium and named after Belgians (although some may only have been Belgian citizens for a limited period of time). In both parts, the minerals are arranged in alphabetical order, except for minerals with a name derived from an already known mineral, such as meta- and para-types, or minerals with varying chemical composition. These are discussed jointly with the normal type. As noted previously, not all minerals treated in the book are included in the official IMA-Mineral database. It was the aim of the authors to keep present-day students, scientists, mineral collectors and nature-lovers also informed about mineral names that, being nowadays discredited or labelled as questionable by IMA, will soon disappear or already faded out from scientific literature and handbooks. As they are part of the Belgian mineralogical heritage and throw light upon the evolution of the mineralogical research in Belgium, they are also included in the present book.

Several handbooks and databases claim that julienite and vaesite are minerals named after Belgian citizens. In-depth inquiries made by the authors of this book provided contradictory data about the nationality of both Henri Julien and Johannes Vaes. Hence, these minerals are not discussed hereafter. Any reader able to provide the authors with decisive data to the matter in hand are kindly requested to inform them.

2. The layout and the content of the **mineral descriptions**, as well as their addenda, require additional comments and clarification.

2.a. The **heading** of each mineral description is marked by a bar displaying the name of the treated mineral. The IMA status of the mineral is instantly deducible from the font style. A mineral name printed in bold capital letters refers to an IMA approved species. Minerals shown in bold italic capitals have a questionable status, whereas those in regular italic capitals have been discredited. When the mineral name was discredited, the approved name of the mineral is also given: in regular capital letters are mineral species that are not described in this book, names in bold capitals refer to minerals described *in extenso*.

The bar with the mineral name is followed by an insert (or **focus box)** that documents basic features of the mineral put in the forefront.

2.b. With regard to the **origin of the mineral name**, which is illustrated at the upper end of the focus box, the reported information is purposely limited to the name of the individual, the locality, a geographical or geological feature, or the chemical composition embodied in the mineral name.

2.c. As the **chemical formula** of a specific mineral may differ a lot, depending on the author or the handbook the reader will consult, it was decided to include here the formula recommended by Nickel, E.H. & Nichols, M.C. (updated version, March 2009) The official IMA-CNMNC List of Mineral Names (http://pubsites.uws.edu.au/ima-cnmnc). This mineral database is provided on the web by Materials Data Inc., Livermore, CA, USA (www.materialsdata.com).

2.d. For evident reasons, data with regard to the **crystal system** doesn't require further comments.

2.e. The **type locality** of a mineral is known to be the specific geographical locality where this mineral was first recognized and described. When reference is made to a bilingual municipality, then the name of the locality is reported in the two languages ruling locally. When the type locality is a hamlet, or a somewhat larger subdivision of a present-day Belgian municipality then, as a rule, the name of this municipality is shown between brackets and put immediately after that of the type locality.

In February 2009 the government of the Democratic Republic of the Congo introduced a new administrative subdivision for the country. Changes with repercussions on the present book primarily concern the former Katanga Province. According to the new subdivision, the former Katanga Province is now split into three new provinces: the Haut-Katanga Province, the Lualaba Province and the Haut-Lomami Province. What formerly was called the Bas-Congo Province is known today by the name of Central Kongo Province. In the parts of this book concerned with mineral descriptions, as a rule, only the original French names have been mentioned (http://democratie.francophonie.org/IMG/Constitution de la RDC.pdf. *See Article 2*).

2.f. The mineralogical definition of the term **type material** refers to all the specimens upon which the description of the new mineral species is based. According to Stalder's report (2002) (www.agiweb.org/smmp/IMA-CM/ctms_report2002.pdf) the term **holotype** (HT) refers to a single specimen (designated by the author) from which all the data for the original description were obtained. **Cotype** (CT) material denotes specimens (designated by the author) used to obtain quantitative data for the original descriptions. A **neotype** (NT) is a specimen chosen by the author of a redefinition or re-examination of a species to represent the species when the holotype or cotypes cannot be found. A **type specimen** (T) is (only) very probably a "real" type specimen (HT or CT) or at least a specimen determined by the author. The catalogue of type mineral species is available as pdf files, listed alphabetically: www.agiweb.org/smmp/IMA-CM/ctms.

2.g. The meaning of the symbols used to denote the **IMA status** of the mineral is as follows: G = grandfathered, first described prior to the foundation of the IMA (1959), Q = questionable, Rd = redefinition approved by the CNMNC, Rn = renaming with the approval of the CNMNC, and D = discredited. A code as for example 2004-18, means that the mineral name was approved by the CNMNC in 2004 and that it was the eighteenth new mineral name approved that year.

2.h. References given at the lower end of the focus box inform the reader on the first description of the mineral and other significant data. Reference is also made to the mineral description as given by the Mineralogical Society of America in the Handbook of Mineralogy, and freely distributed as a pdf file to the public on the website of the MSA (www.handbookofmineralogy.org/). When reference is made to Belgian journals with a bilingual title then, as a rule and to save space, the referred title (usually either in Dutch, or in French) is indicative of the language used by the author of the article.

2.i. All displayed **crystal drawings** are adapted drawings from literature. Miller indices and the current orientation of the crystals replace old indices and the original orientation. SHAPE software (version 6) for Windows (copyright Eric Dowty, 1999), put at the disposal of the authors by Mineralcolor, was used to adapt drawings published previously.

Focus boxes may be followed by **additional comments or remarks** related to the history or status of the mineral.

3. Next, to help the reader visualize the mineral, one or several **photographs** are incorporated.

As the idea to write this book dates from many years before the advent of digital photography, all photographs illustrating the text are scans of slides taken with Fuji Sensia films 100 ASA, using an Olympus OM-2 camera equipped with macro objectives with the following focal distances: 90 mm and 50 mm, or 38 mm and 20 mm, mounted on a bellow. As illumination source a of halogen was used light conducted through glass fibres (Schott KL 1500-2) and converted into daylight with a daylight filter. All photographs, with the exception of the SEM and TEM photographs (SEM = Scanning electron microscope; TEM: transmission electron microscope), were taken by the first author (E. Van Der Meersche). Photographs put at the disposal of the authors are easily identifiable because their legend explicitly mentions their origins.

The width of the photographed field (enlargement or scale) is mentioned in one of the corners of each photograph.
Photograph captions give a brief description of the specimen, the name of the locality or the site where it came from, and the collection. Country names are usually abbreviated (B = Belgium; D.R. Congo = Democratic Republic of the Congo; F = France).

4. The **biographical note** which accompanies the focus box and the photographs is purposely concise. It provides aspects worth knowing about the life, education, professional career and achievements of the individual honoured with the mineral name. Only a few biographical notes are poorly documented. Readers disposing of relevant data about less documented individuals are therefore invited to contact the authors and to pass on their information in view of a hypothetical new edition of the present book.

The terminology and abbreviations used in the biographical text, when reference is made to Belgian universities, ask for some explanation. As a rule, the universities established in Ghent and Liège are designated as Ghent University (GU) and University of Liège (ULg), respectively. When reference is made to one of the universities in Brussels - for convenience and to avoid confusion among readers not familiar with the Belgian linguistic cauldron - following terms and abbreviations have been used. The term Brussels Free University (BFU) denotes the common ancestor institution of the two entities or sister institutions officially created in the early seventies of last century and known today as the French-speaking *Université Libre de Bruxelles* (ULB), and the Dutch-speaking *Vrije Universiteit Brussel* (VUB). The university founded in Leuven a long time ago was split at about the same time as the BFU to form two distinct entities or sister universities. In the present work, the historic or 'old' university, from the early beginning located in Leuven, is referred to as the Catholic University of Leuven (CUL). After the split, the Dutch-speaking *Katholieke Universiteit Leuven* (KUL) remained in Leuven, whereas the French-speaking *Université catholique de Louvain* (UCL), moved to Louvain-la-Neuve (Ottignies). The authors are aware of the fact that the terms and abbreviations mentionned above are not always in agreement with those postulated by official Belgian organisms but at least they have the advantage of being clear-cut and understandable for people from abroad.

The translation into English of scientific and academic ranks used in the past and today in Dutch-speaking and French-speaking Belgian universities was sometimes an unbridgeable problem. When no appropriate English term was available the authors gave preference to use the French terminology as, both in the Dutch- and in the French-speaking parts of Belgium, it was in current use in the 19th and early 20th century. An identical problem emerged when they had to translate the different ranks used years ago in the administration of private companies and governmental societies in Belgium. In these cases the French nomenclature was used, the more that many French terms are easily understood by non French-speaking people.

A **portrait** of the documented person (when available) has been added to the biographical note. The majority of the portraits illustrating the biographical notes was borrowed from earlier publications. Still living individuals have been asked to send a recent photograph of their person to the authors. They always complied with their request. The names of all persons, academies, universities and other institutions that authorized to publish photographs that passed into their ownership are listed at the end of the book (*see*: **Portrait Credits**). They are greatly acknowledged.

Biographical and bibliographical references associated with the biographical note, when available, provide interested readers with more details. Their number was kept limited to save space and to avoid double use. They are listed chronologically and were selected arbitrarily to some degree. When a biographical or a bibliographical reference makes mention of a bilingual publication, then, in accordance with what was mentionned above, the quoted title of the publication informs the reader on the language used by the author of the article.

5. Private persons, scientists, institutes, societies and mineral clubs that provided information of any kind that helped the authors accomplish far more than they could have done alone are reported at the end of the book (**Acknowledgements**). **Portrait Credits** are mentioned next. These two appendices are followed by a **Personal Names Index** and a **Mineral Names Index**, both listed alphabetically.

PART I. MINERALS DISCOVERED IN BELGIUM

ARDENNITE-(As)

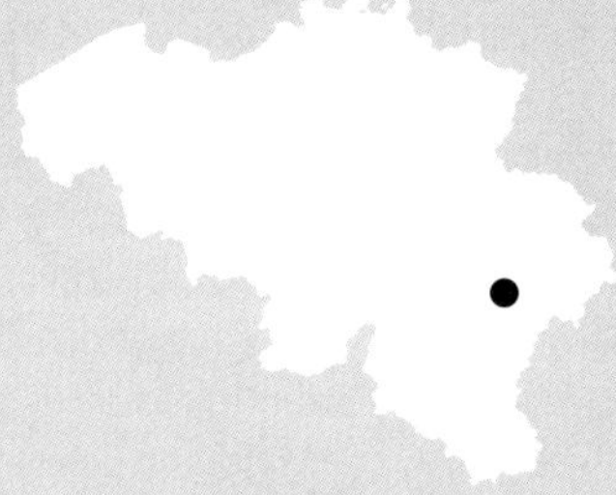

The Ardennes, a region in southern Belgium and neighbouring countries.

$(Mn^{2+},Ca)_4(Al,Mg,Mn^{3+})_6(AsO_4)(SiO_4)_2(Si_3O_{10})(OH,O)_6$

Orthorhombic.

(?) Ottré (Vielsalm), Luxembourg Province, Belgium.

Poland: Mineralogical Museum University of Wroclaw - Type: CT: MMWr-14794 (donated by von Lasaulx) and MMWr14796 (donated by Bettendorf).
France: Musée de Minéralogie, Ecole des Mines de Paris - Type: *dewalquite* [donated by Pisani (with label). Pisani obtained this specimen from the Adam collection (N°2241) in 1873]. The name *dewalquite* was discredited, the mineral being the same as ardennite-(As).

IMA: Rn (publication pending). The suffix -(As) was added after the discovery of ardennite-(V).

von Lasaulx, A. (1872a) Ardennit ein neues Mineral. *Verhandlungen des naturhistorischen Vereines der preussischen Rheinlande und Westphalens*, **29**, 174.

von Lasaulx, A. (1872b) Über ein neues Mineral aus der Gegend von Ottrez. *Verhandlungen des naturhistorischen Vereines der preussischen Rheinlande und Westphalens*, **29**, 189-192.

Pisani, F. (1872) Sur un nouveau silico-aluminate de manganèse vanadifère trouvé à Salm-Château, en Belgique. *Comptes rendus de l'Académie des Sciences, Paris,* **75**, 1542-1544.

von Lasaulx, A. & Bettendorf, A. (1873) Ardennit, ein neues Mineral. *Annalen der Physik und Chemie*, **149**, 241-251.

Pisani, F. (1873) Analyse de la dewalquite de Salm-Château, en Belgique. *Comptes rendus de l'Académie des Sciences, Paris,* **77**, 329-333.

von Lasaulx, A. (1876) Nachträge zur Kenntniss des Ardennites. *Neues Jahrbuch für Mineralogie,* **1876,** 363-368.

Buttgenbach, H. (1931-32) Sur un cristal de dewalquite. *Annales de la Société géologique de Belgique*, **50**, B176.

Barresi, A., Orlandi, P. & Pasero, M. (2007) History of ardennite and the new mineral ardennite-(V). *European Journal of Mineralogy*, **19,** 581-587.

http://www.handbookofmineralogy.org/pdfs/ardennite.pdf

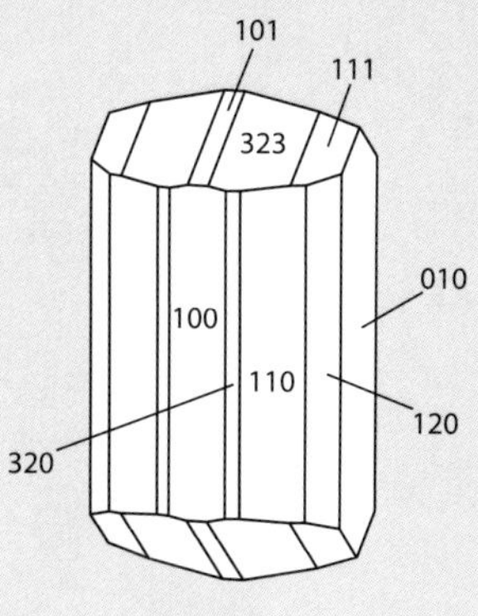

after von Lasaulx & Bettendorf (1873)

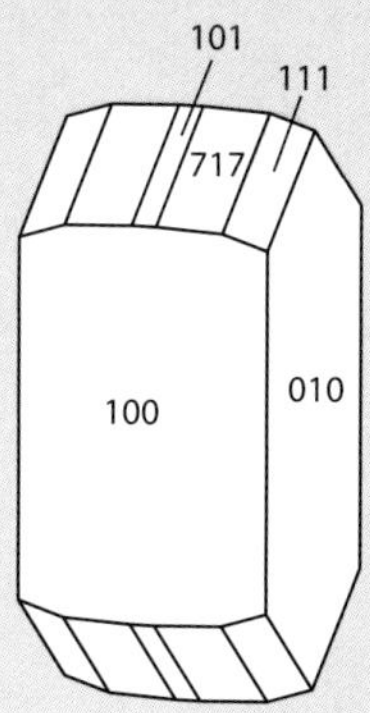

after Buttgenbach (1931-32)

Ardennite was described as a new mineral species by the German mineralogist and petrographer Arnold von Lasaulx (von Lasaulx, 1872a and 1872b) and von Lasaulx & Bettendorf (1873). As von Lasaulx bought the specimens, the origin of the material he analysed remains uncertain though he claimed that the sample came from the surroundings of Ottré. According to von Lasaulx, ardennite was a variety of kyanite. This explains why in some older articles the mineral is designated as *manganese kyanite*.
At about the same time as von Lasaulx, but completely independent of him, the same mineral was also described by Félix Pisani, a French chemist and mineral dealer (Pisani, 1872 and 1873). The sample Pisani analysed came from Salm-Château (Vielsalm). He gave it the name *dewalquite* in honour of Gustave Dewalque (*see*: *dewalquite*). Chemical analyses of the material examined by Pisani pointed out the relatively high vanadium and arsenic contents of the mineral.
According to IMA the mineral should be named ardennite, as proposed by von Lasaulx.
After the discovery of ardennite-(V) in 2007, the original name ardennite was modified into ardennite-(As).

SEM photographs of ardennite-(As) crystals. Salmchâteau (Vielsalm), Luxembourg Province, B. Royal Belgian Institute of Natural Sciences.

*Aggregate of ardennite-(As) in quartz.
Bihain (Vielsalm), Luxembourg Province, B.
M. Blondieau.*

*Prismatic crystal of ardennite-(As).
Salmchâteau (Vielsalm), Luxembourg Province, B.
J. Detaille.*

*Group of ardennite-(As) crystals in quartz cavity.
Salmchâteau (Vielsalm), Luxembourg Province, B.
M. Blondieau.*

*Acicular crystal of ardennite-(As).
Bihain (Vielsalm),
Luxembourg Province, B.
M. Blondieau.*

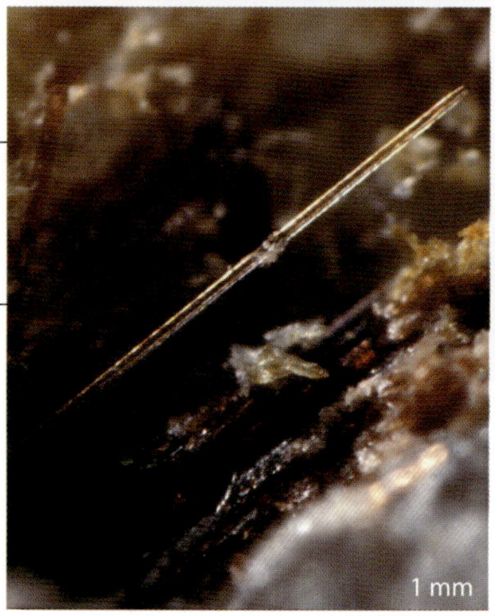

*Tabular crystal of ardennite-(As).
Salmchâteau (Vielsalm), Luxembourg Province, B.
Laboratory of Mineralogy, University of Liège.*

ARDENNITE-(V)

The vanadium-dominant analogue of **ardennite-(As)**.

$(Mn^{2+})_4(AlMg)Al_4(Si_5V)O_{22}(OH)_6$

Orthorhombic.

Sparone, Locana valley, Torino Province, Italy.

Italy:
Centro Interdipartimentale Museo di Storia Naturale e del Territorio,
University Pisa-Calci (PI) - Type: HT no. 15525.
Museo Regionale di Scienze Naturali, Torino - Type: CT N° M/U 15544.

IMA N°: 2005-037

Barresi, A., Orlandi, P. & Pasero, M. (2007) History of ardennite and the new mineral ardennite-(V). *European Journal of Mineralogy*, **19,** 581-587.

The discovery of ardennite-(V) by Barresi *et al.* (2007) led to rename *ardennite* into ardennite-(As). These authors describe in detail the competition between von Lasaulx and Pisani about the discovery of the mineral, and their successive analyses.
Ardennite-(V) has so far not been found in the Belgian Ardennes.

BASTONITE -> ammonium-bearing variety of BIOTITE

'Les Roches' quarry, Bastogne, Luxembourg Province, Belgium.

IMA: D

Dumont, A. (1848) Mémoire sur les terrains ardennais et rhénan de l'Ardenne, du Rhin, du Brabant et du Condroz, *Mémoire de l'Académie royale de Belgique*, 2 vol., **20** and **22**, 451 p.

Renard, A. (1882) Les roches grenatifères et amphiboliques de la région de Bastogne. *Bulletin du Musée royal d'Histoire naturelle de Belgique,* **1**, 38-43.

Klement, C. (1888) Analyses chimiques de quelques minéraux et roches de Belgique et de l'Ardenne française. *Bulletin du Musée royal d'Histoire naturelle de Belgique*, **5**, 159-186.

Prinz, W. (1909) Les micas et filons granitoïdes de Bastogne. *Bulletin de la Société belge de Géologie*, **23**, 129-132.

Bastonite was discovered in the 'Les Roches' quarry by André Dumont (*see*: *dumontite*). At this place, the country rock is a quartzite. In his publication, Dumont (1848) often mentions the term *bastonifère*, without referring to the origin of *bastonite*.
Renard (1882) presents the results of his observations and analyses, and concludes that *bastonite* has many properties in common with phlogopite.
Klement (1888) (*see*: *klementite*) provides the first chemical analysis of *bastonite*.
Prinz (1909) refers to the French geologist Jules Gosselet who presumed that *bastonite* was a variety of biotite. Morphological and optical data confirmed this supposition.

BELGITE -> **WILLEMITE**

Refers (erroneously) to Belgium.

Altenberg/Vieille-Montagne, Kelmis/La Calamine, Liège Province, Belgium.

IMA: D

Lévy, A. (1830) Lettre à l'éditeur. *Neues Jahrbuch für Mineralogie,* **1**, 71.
Panebianco, R. (1916) Belgito. *Rivista di Mineralogia e Cristallografia Italiana,* **47**,13-15. (in Esperanto).
Spencer, L.J. (1919) Eighth list of New Mineral Names, *Mineralogical Magazine*, **18**, 373-390.

In 1916 the Italian scientist Ruggero Panebianco proposed to give up the use of the name willemite (*see*: willemite), introduced by Armand Lévy in1830. According to Panebianco it was inappropriate to name a new mineral species after a king, in this case King William I (Willem I in Dutch) of the Netherlands, and he suggested to use the name belgite from then onward, after the country where, according to Panebianco, the first mineral specimens came from. At the time Panebianco made his proposal, however, the quarry of Altenberg/Vieille-Montagne, presently part of Kelmis/La Calamine, in the Liège Province, did not belong (*see*: below) to the Kingdom of Belgium. Thus, not surprising in later times the name belgite was discredited and the older name willemite rehabilitated.

A historical insertion:

After the fall of Napoleon (1815), the borders between the United Kingdom of the Netherlands and the Kingdom of Prussia were re-established. At Moresnet, however, on account of the great economic interest of the zinc ore deposits of Kelmis, both nations could not come to an agreement and the territory of the municipality was divided into three parts. It was decided (the *Borders Treaty*, or *Aachen Border Treaty*) that the place where the zinc ore was quarried should become a neutral territory, called Neutral Moresnet.
The village of Moresnet itself, located west of Neutral Moresnet, was allocated to the Netherlands, whereas Neu-Moresnet, to the east, became part of Prussia.
Neutral Moresnet, with a surface area of hardly 3.5 km^2 and a median position, had numerous attributes of a sovereign state (anthem, flag and stamps) and was jointly ruled by Dutch and Prussian commissioners. The existence of this mini-state lasted over one century (from 1816 to 1919). In 1908, Wilhelm Molly, a German general practitioner, proposed that Neutral Moresnet should become the world's first Esperanto-speaking state, named *Amikejo* ('place of friendship').

At the end of World War I, as a result of the *Treaty of Versailles*, Neutral Moresnet was awarded to the Kingdom of Belgium. Presently, the depleted mining area of Moresnet belongs to Kelmis (La Calamine), a German-speaking municipality in the north-eastern part of the Liège province.

CALAMINE -> a mixture of non-sulphide zinc minerals such as SMITHSONITE, HEMIMORPHITE, HYDROZINKITE and/or **WILLEMITE.**

Kelmis/La Calamine, Liège Province, Belgium, is named after the zinc ore *calamine*.

IMA: D

Davreux, C.J. (1833) Essai sur la constitution géognostique de la province de Liège. *Mémoires couronnées de l'Académie royale des Sciences et Belles-Lettres de Bruxelles*, **9**, 297 p.

Lévy, A. (1843) Description de plusieurs espèces minérales appartenant à la famille du zinc. *Annales des Mines*, Paris, 4e série, **4**, 517.

Cesàro, G. (1886) Note sur une nouvelle face de la calamine. *Bulletin de la Société française de Minéralogie*, **9**, 242.

Buttgenbach, H. (1947) Les minéraux de Belgique et du Congo Belge. 573 p., Paris: Dunod.

Blondieau, M. & Polrot, F. (2004) Métallurgie ancienne. Petite note sur un déchet stratifié métallique lourd. *Bulletin de la Société royale belge d'Etudes géologiques et archéologiques. « Les Chercheurs de la Wallonie »*, **43**, 67-70.

Dejonghe, L. & Boni, M. (2005) The "calamine-type" zinc-lead deposits in Belgium and West Germany: a product of Mesozoic palaeoweathering processes, *Geologica Belgica*, **8**, N° 3, 3-14.

http://kelmis.be/gemeindeportrait/geschichte (October 2009)

The name of La Calamine (Kelmis in Dutch), where since Roman times until the end of the 19th century zinc ore was extracted, refers to *calamine*, or zinc ores.
Davreux (1833) refers to Plinius the Elder, who mentioned in his book 34, chapter 1, that a large *cadmia* deposit was discovered and exploited in the vicinity of modern Kelmis.
Theophilius (12th century) uses the name *calamine* to describe the quarried material.
Blondieau & Polrot (2004) point out that *cadmia* and *calamine*, two old miner terms used as synonyms, do not have the same meaning, *calamine* being the ore, and *cadmia* the melting product or a kind of slag material.

Today the municipality of Kelmis/La Calamine forms part of the German-speaking region located in eastern Belgium.
The use of the name 'Kelms' is thought to go back to 1280 and to derive from the Latin term *cal(a)mis*, meaning *near to the calamina* (zinc ore).
The French name 'Calamine' was probably mentioned for the first time on the Martini map (http://kelmis.be/gemeindeportrait/geschichte, October 2009).

In early Belgian handbooks and scientific papers on mineralogy, prominent mineralogists such as Lévy (1843), Cesàro (1886) and Buttgenbach (1947) assume that *calamine* is a mineral species. In more recent times, however, it became obvious that calamine is not a single mineral species but a mixture of different zinc minerals, such as smithsonite, hemimorphite, hydrozinkite and/or willemite (*see:* willemite).

Dejonghe & Boni (2005) use the term *calamine* to refer to "non-sulphide zinc ores of supergene origin in carbonate rocks".

CALCLACITE

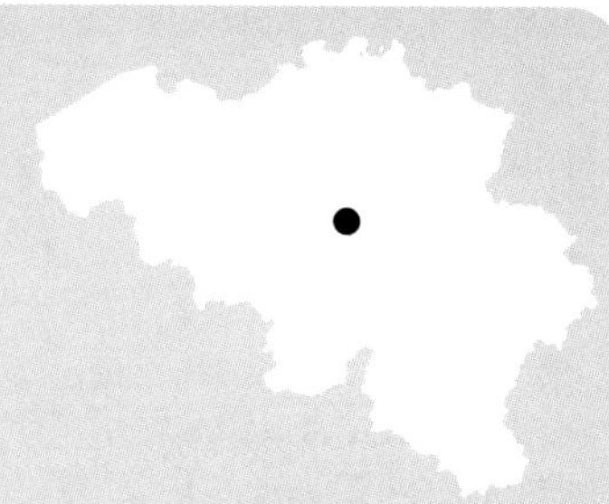

Chemical composition: **Cal**(cium), **C**(h)**l**(orine), and **Ac**(etate).

$Ca(CH_3COO)Cl.5H_2O$

Monoclinic/Triclinic.

Royal Belgian Institute of Natural Sciences, Brussels, Belgium.

Belgium: Royal Belgian Institute of Natural Sciences, Brussels - Type: HT/CT: RN 5518 (many samples).

IMA: G

Taboury, F. (1931) Des modifications chimiques de certaines substances calcaires conservées dans des meubles en bois. *Bulletin de la Société chimique de France*, **49**, 1289-1291.

Van Tassel, R. (1945) Une efflorescence d'acétatochlorure de calcium sur des roches calcaires dans des collections. *Bulletin du Musée royal d'Histoire naturelle de Belgique,* **21**, 1-11.

http://www.handbookofmineralogy.org/pdfs/calclacite.pdf.

René Van Tassel (*see*: vantasselite), who discovered and described this mineral, found it to form white efflorescences on limestone fragments and fossils stored in the collections of the Royal Belgian Institute of Natural Sciences in Brussels. It is most remarkable that in this context the mineral apparently only crystallizes on specimens stored in drawers made of oak. As far as we know, calclacite does not occur in nature.
Crystals of similar composition were described over a decade earlier by Taboury (1931) on the occasion of a study of efflorescences noticed on carbonate substances stored in oaken pieces of furniture at the Faculty of Sciences of the University of Poitiers (France). According to Félix Taboury, the efflorescences consisted of a simple acetate of calcium, $Ca(CH_3COO).2H_2O$, although no chemical analysis was provided by him to substantiate his conclusion.

Acicular calclacite on limestone from Yvoir, Namur Province, B., stored in oaken cabinets at the Royal Royal Belgian Institute of Natural Sciences. Holotype (N° N 5518).

Carl Camerman (1885-1958).

Sint-Amands, Antwerp Province, Belgium.

IMA: D

Denaeyer, M.E. & Ledent, D. (1952) Sur la présence de la modification hexagonale de la hiératite (*camermanite*) dans des incrustations de sels potassiques d'une cheminée d'usine. *Bulletin de la Société française de Minéralogie et de Cristallographie*, **75**, 231-236.

Gramaccioli, C.M. & Campostrini, I. (2007) Demartinite, a new polymorph of K_2SiF_6 from La Fossa crater, Vulcano, Aeolian Islands, Italy. *The Canadian Mineralogist*, **45**, 1275-1280.

Denaeyer and Ledent of the Laboratory of Mineralogy of the Brussels Free University - two colleagues of Carl Camerman - proposed the name *camermanite* for this hexagonal form of K_2SiF_6, dimorpous with cubic hiératite.
The material was found by Camerman at Sint-Amands, in the province of Antwerp. Here, it occurred in incrustations of potassium salts on bricks from a factory chimney, used in former days for venting industrial gases (HF, H_2SO_4,SO_2, etc.) but presently demolished.
As the origin of this material was artificial, *camermanite* was not accepted as a mineral.

Recently, natural hexagonal K_2SiF_6 has been described as demartinite from a fumarole in the La Fossa crater, Vulcano, Aeolian Islands, Italy (Gramaccioli & Campostrini, 2007).

Carl Camerman was born at Leuze (Leuze-en-Hainaut), in the Hainaut Province, on November 15, 1885. In 1903, he enrolled as a pre-engineering student at Ghent University. Upon the completion of his second year of engineering studies, he left for the Brussels Free University (BFU), where he graduated as a mining engineer in 1908. Early on in his professional career, Camerman served as a chemist in a laboratory for material testing administered by the Belgian Railway Company in Mechelen and headed by his father. Here he became thoroughly acquainted with the chemistry of building-materials.

Two years after World War I, Camerman accepted a position as a geologist with the oil company *Petrofina* (1920). In this capacity, he travelled to the USA, the Middle East, North Africa and Romania. Next, he became manager of the *Asphalt Block Pavement Company*, a post he kept until the onset of the Second World War. In this period Camerman also headed the laboratory of the *Office de Recherches et d'Expérimentation* (OREX) in Brussels. All these activities yielded an extensive knowledge of the chemical and physical properties of stony materials, bricks and clays used in construction. No wonder that many industries and societies involved in building and building-materials appealed to his professional experience.

During World War II, Camerman could not continue these above-mentioned activities. Instead he set himself out to study in a systematic way limestone outcrops in the neighbourhood of Tournai and in other regions of central and southern Belgium. Meanwhile, he worked on his own account as a chemist specialized in the study and analysis of building-materials and water. After the war, he resumed his collaboration with OREX and concurrently worked with Louis Baes, who lectured at the BFU and the Brussels Academy of Fine Arts. Baes entrusted him with a course dealing with engineering materials used in the building-trade, problems related to efflorescence formation, and brick and tile production. Courses treating the same subjects to be lectured at the *Institut Supérieur d'Architecture de La Cambre* were also committed to his charge.

Camerman joined the Belgian Geological Survey in Brussels in 1948. Here, his scientific work focused on various topics, including (a) weathering and other deterioration processes of building stones, mortars and bricks as a result of industrial and atmospheric pollution, (b) efflorescence mechanisms affecting stony materials and ceramics, and (c) techniques of conservation, cleaning and waterproofing of rocks such as e.g. the Middle Eocene Lede (or Balegem) sandstone. The latter rock type was widely used e.g. in Gothic and Renaissance religious and public buildings and monuments in Brussels, Flanders, the Walloon region and the southern Netherlands. Camerman also kept an eye on the ground and building works rea-lized in Brussels in view of the World Fair 1958. He died in his country house at Biez (Grez-Doiceau), in the Walloon Brabant Province, on November 1, 1958.

Biography

de Roubaix, J. (1959) Eloge funèbre de Carl Camerman prononcé le mardi 16 décembre à la Société Belge de Géologie, de Paléontologie et d'Hydrologie. *Bulletin de la Société belge de Géologie, de Paléontologie et d'Hydrologie,* **67**, 521-524.

Bibliography

Camerman, C. (1960-61) Les pierres naturelles de construction, *Annales des Travaux publics de Belgique*, **4**, 48-51.

CIPLYTE -> APATITE

Named for the bed (stratum) known as "phosphatized chalk of Ciply".

Ciply (Mons), Hainaut Province, Belgium.

IMA: D

Ortlieb, J. (1888) Sur la ciplyte. *Annales de la Société géologique du Nord*, **XVI**, 270-278.
Van Tassel R. (1972) L'apatite de Ciply, Hainaut: une mise au point. *Bulletin de l'Institut royal des Sciences naturelles de Belgique*, **48**, 4, 1-3.

In a letter to Jules Gosselet, the Belgian geologist Jean Ortlieb presents the analyses of brown grains he extracted from the phosphatized chalk of Ciply and concludes they consist of a new mineral species. The latter is now considered to be apatite (Ortlieb, 1888).
In several references, the name Ortlieb is erroneously mentioned as *Ortlier*.
Van Tassel (1972) proves that the specimen labelled as *"apatite de Ciply"* and stored in the collection of the Royal Belgian Institute of Natural Sciences in Brussels, does not come from Ciply but originates from Jumilla in Spain.

DAVREUXITE

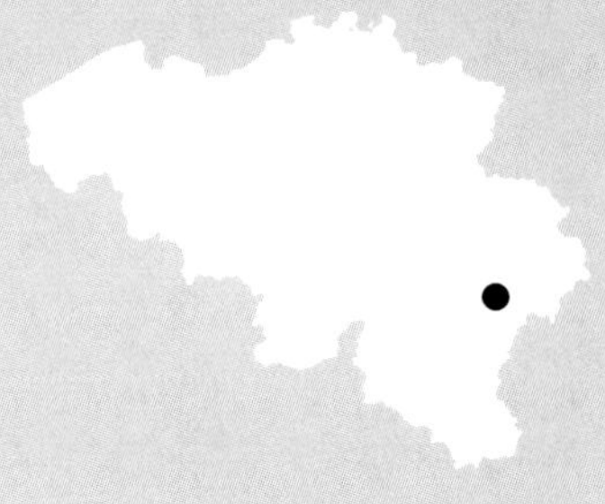

Charles-Joseph Davreux (1800-1863).

$Mn^{2+}Al_6Si_4O_{17}(OH)_2$

Monoclinic.

Ottré (Vielsalm), Luxembourg Province, Belgium.

Belgium: Royal Belgian Institute of Natural Sciences, Brussels - HT: RA 8606.
Poland: Mineralogical Museum University of Wroclaw - Type: CT: MMWr-11504/5 (donated by de Koninck).

IMA: G

Dumont, A. (1852) Exposé sur la situation du Royaume. *Statistique générale de la Belgique, périodes décennales de 1841 à 1850.* Bruxelles: Lesigne.

de Koninck, L.-G. (1878) Sur la davreuxite, espèce nouvelle recueillie dans les filons de quartz du terrain ardennais. *Bulletin de l'Académie royale de Belgique*, 2e série, **46**, 240-245.

Lacroix, A. (1886) Propriétés optiques de quelques minéraux (wavellite, variscite, planérite, davreuxite). *Bulletin de la Société française de Minéralogie et de Cristallographie,* **9**, 4-6.

Buttgenbach, H. (1947) Les minéraux de Belgique et du Congo belge. 573 p., Paris: Dunod.

Fransolet, A.-M. & Bourguignon, P. (1976) Précisions minéralogiques sur la davreuxite. *Comptes rendus de l'Académie des Sciences*, Paris, **283 D**, 295-297.

Strunz, H. (1977) Mineralogische Tabellen. 6.Auflage, 621 p., Leipzig: Akademische Verlagsgesellschaft Geest & Portig.

Fransolet, A.-M., Abraham, K. & Sahl, K. (1984) Davreuxite: a reinvestigation. *The American Mineralogist*, **69**, 777-782.

http://www.handbookofmineralogy.org/pdfs/davreuxite.pdf

Davreuxite was discovered by André Dumont (1852) in quartz veins near Ottré. However, he named the mineral *trémolite fibreuse*.
The same mineral was found at Salmchâteau and based on chemical analyses, de Koninck (1878) proposed to name this new mineral davreuxite.
Lacroix (1886) introduced doubts by mentioning that this mineral was a mica. This assertion was taken over by Buttgenbach (1947) and Strunz (1977), who classified it as a *variety of manganiferous vermiculite*.
Fransolet *et al.* (1984) showed that davreuxite is a valid new mineral species.

Fibrous davreuxite embedded in quartz, Bihain (Vielsalm), Luxembourg Province, B. M. Fouassin.

Charles-Joseph Davreux was born in Liège on September 10, 1800. Following the completion of his secondary schooling at the Imperial Lyceum of his hometown, he decided to become seminarian. The death of his father, a practising pharmaceutical chemist, put his plans for the future on hold. To meet the financial needs of his family, he asked and got permission to temporarily run the drugstore of his departed father. Next, he took courses at the university. During this period Davreux was appointed *préparateur-adjoint* to the course of chemistry lectured by Jean Charles Delvaux (*see*: delvauxite). Thanks to this occupation, he acquired a taste for chemistry and mineralogy. In 1825, he took the degree of pharmaceutical chemist delivered by the provincial medical commission. A few years earlier he already cofounded the *Société des Sciences naturelles de Liège* and in 1827 also contributed to the foundation of the *Ecole industrielle de Liège*.

In 1830, Davreux participated in a prize competition organized by the Royal Academy for Science and Literature in Brussels in order to encourage scientists to study the geological constitution of several Belgian provinces, including those of Brabant, Hainaut, Liège, Luxembourg and Namur. Davreux's memoir on the Liège Province secured the second place, whilst the first prize was awarded to the memoir submitted by André Dumont (*see*: dumontite). Shortly afterwards, Davreux was appointed curator of the mineral collections of the University of Liège, a position he kept until 1834. At the same time he served as *préparateur* to the laboratory of pharmaceutical chemistry headed by Henri Gaëde. When the latter passed away, Davreux accepted to teach mineralogy and crystallography. In 1836 he was no longer interested in an academic career. He preferred to devote himself to more practical activities in the fields of industrial chemistry and toxicology. Davreux joined the medical commission of the Liège Province in 1840 and was appointed adjunct member and title-holding member of the Royal Academy for Medicine of Belgium in 1842 and 1854, respectively. In the meantime he was nominated *agrégé* to the Liège alma mater (1845) and over the years became increasingly interested in local archaeology and history. Davreux died in Liège on April 11, 1863.

Biography

Dewalque, G. (1873) *Biographie nationale*, **4**, col. 733-735.

Bibliography

Dewalque, G. (1873) *Biographie nationale*, **4**, col. 735.

DELVAUXITE

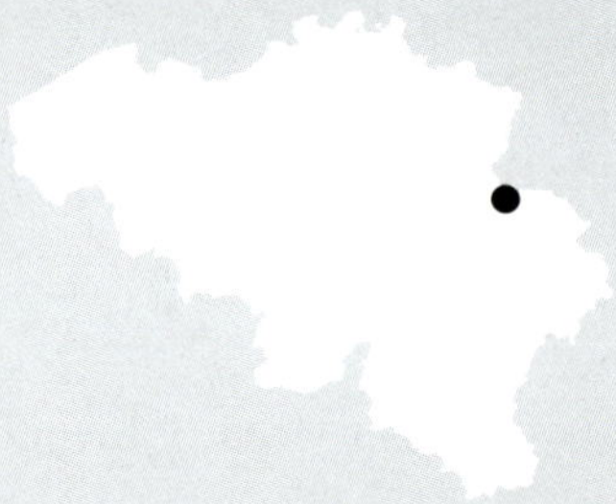

Jean Charles Delvaux de Fenffe (1782-1863).

$Ca(Fe^{3+})_4(PO_4)_2(OH)_8.4\text{-}5H_2O$

Amorphous.

Berneau (Dalhem), Liège Province, Belgium.

France: Muséum national d'Histoire naturelle, Paris - Type: T: 38.37 (donated by A. Dumont; small fragments).

IMA: Q

Dumont, A.-H. (1838) Notice sur une nouvelle espèce de phosphate ferrique. *Bulletin de l' Académie royale de Belgique*, **5**, 296-300.

Čech, F. & Povondra, P. (1979) A re-examination of borickýite. *Tschermaks Mineralogische und Petrografische Mitteilungen*, **26**, 79-86.

Fleischer, M., Cabri, L.J., Chao, G.Y. & Pabst A. (1980) New Mineral Names. *The American Mineralogist,* **65**, 807-814.

http://www.handbookofmineralogy.org/pdfs/delvauxite.pdf

In 1793, André Dumont's father collected brown resinous specimens at Berneau, which Dumont Jr. named lithomarge. After collecting more samples himself, Dumont asked Delvaux to analyse the material. The chemical composition was different from that of similar material, so that Dumont (1838) concluded that he had found a new mineral, which he named delvauxine in honour of Delvaux.

Fleischer *et al.* (1980) refer in the section "New Data" to an article of Čech & Povondra (1979) in which the authors claim that delvauxite, borickýite and foucherite have the same characteristics. As delvauxite was already described in 1838, this name had priority and the two other names were discredited, but re-examination of type material is required to get a clear status.

Chestnut-coloured delvauxite. Richelle (Visé), Liège Province, B. J. Van Daele.

Crackled delvauxite crust.
Richelle (Visé), Liège Province, B.
B. Darimont.

Jean Charles Philippe Joseph **Delvaux de Fenffe** was born in Rochefort, in the Namur Province, on July 25, 1782, son of Théodore Joseph, a physician. Jean Charles received his early education in Liège, where a maternal uncle was canon. At the age of 18, he attended the Faculty of Medicine in Paris, and graduated as a doctor in medicine early-on in 1806. On his return to Belgium, he took over part of the practice of his father. As this job did not live up to his expectations, he left Rochefort and settled in Liège in 1809. Here, he became a professor of physical sciences at the Imperial Lyceum (1810), yet without completely renouncing his medical practice. In 1811, Delvaux joined the Faculty of Sciences of the Imperial University of Liège, which had been founded shortly before. It was his responsibility to lecture physics and chemistry. At about the same time he was appointed an officer to the university. In 1812, he became a member of the academic board and graduated as a doctor in sciences.

When in 1817, through King William I of The Netherlands, the University of Liège was founded Delvaux chaired the Faculty of Sciences which at the time consisted of two professors only (Delvaux and Jean Michel Vanderheyden). In this capacity he had to teach experimental physics, general chemistry, chemistry applied to the arts, and metallurgy. During the academic year 1832-1833 he was elected rector and between 1819 and 1837 he occupied the position of Dean of the Faculty of Sciences several times. He retired from the university in 1837, aged 55. After retirement, he devoted himself to personal affairs and continued his medical practice. In 1857, Delvaux withdrew to the castle of Fenffe (Houyet), in the Namur Province, where he died on November 14, 1863, more than 80 years old.

Biography

Delvaux de Fenffe, A. (1925) La Seigneurie de Fenffe et de Hérock, 52-56. Tongres: George Michiels-Broeders.

Florkin, M. (1963) *Annuaire de l'Académie royale de Belgique*, **CXXIX,** 3-21.

Nagant de Deuxchaisnes, A. (2002) La famille Delvaux à Rochefort et à Fenffe. Cercle Culturel et Historique de Rochefort, Cahier n° **57**, 11-26.

Bibliography

Some chemical analyses made by Delvaux are included in papers and books written by Richard Courtois, André Dumont, Charles Davreux and Toussaint.-Dieudonné Sauveur.

DESTINEZITE

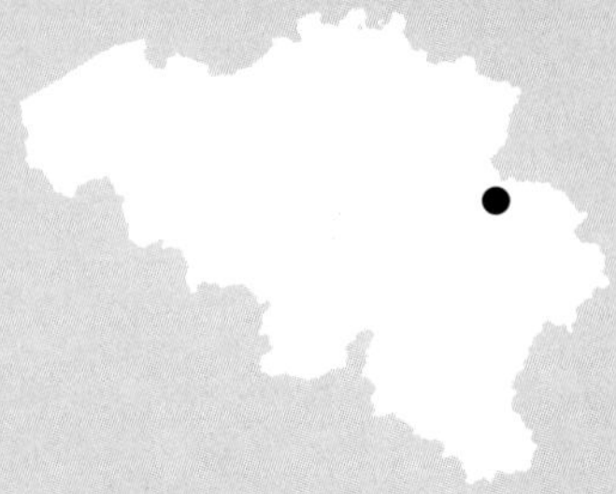

Pierre Destinez (?-1911).

$(Fe^{3+})_2(PO_4)(SO_4)(OH).6H_2O$

Triclinic.

Argenteau (Visé), Liège Province, Belgium.

IMA: Rd (00-E)

Dewalque, F. (1879) Note sur un échantillon de diadochite de la mine de Védrin. *Annales de la Société géologique de Belgique*, **7**, 112-115.

Forir, H. (1879) Sur quelques minéraux et roches trouvés dans une excursion à Argenteau. *Annales de la Société géologique de Belgique*, **7**, 115-118.

Jorissen, A. (1880) Note sur la diadochite (destinézite) de Visé. *Annales de la Société géologique de Belgique*, **8**, 66-69.

Cesàro, G. (1884) Etude chimique et cristallographique de la destinézite. *Annales de la Société géologique de Belgique*, **12**, 173-191.

Van Tassel, R. (1985) Minéraux phosphates secondaires (vashgeyite, destinézite [diadochite], wavellite, crandallite, phosphate de fer) à Haut-le-Wastia, Province de Namur (Belgique). *Bulletin de la Société belge de Géologie*, **94**, 19-27.

Peacor, D.R., Rouse, R.C., Coskren, T.D. & Essene, E.J. (1999) Destinezite ('diadochite'), $Fe_2(PO_4)(SO_4)(OH)\cdot 6H_2O$: its crystal structure and role as a soil mineral at Alum Cave Bluff, Tennessee. *Clays and Clay Minerals*, **47**, 1-11.

Grice, J.D. & Ferraris, G. (2003) New minerals approved in 2002 and nomenclature modifications approved in 1998-2002 by the Commission New Minerals and Mineral Names (CNMMN), International Mineralogical Association (IMA), *The Canadian Mineralogist*, **41**, 795-802.

Destinezite is not mentioned in the Handbook of Mineralogy.

In the same volume of the *Annales de la Société géologique de Belgique* Forir (1879) and Dewalque (1879) reported the occurrence of two new mineral species, destinezite and diadochite, respectively. Over a century later, Peacor *et al.* (1999) demonstrated that the mineral analysed by Dewalque was also destinezite.
For many years there was a lot of confusion about destinezite and diadochite. Their names were often used as synonyms (Jorissen, 1880; Van Tassel, 1985).
Peacor *et al.* (1999) proposed to use the name destinezite for crystalline $(Fe^{3+})_2(PO_4)(SO_4)$ $(OH).6H_2O$, and diadochite for massive to earthy, amorphous material with a chemical composition slightly different from that of destinezite. The proposal was accepted by the IMA-CNMMN (Grice & Ferraris, 2003). The same proposal was already made by Cesàro as early as 1884!

Cream-coloured crust of destinezite.
Haut-le-Wastia (Anhée), Namur Province, B.
Laboratory of Mineralogy, University of Liège (N° 20016).

Tuberiform destinezite on shale.
Blaton (Bernissart), Hainaut Province, B.
Royal Belgian Institute of Natural Sciences (N° 3321).

Tuberiform destinezite.
Richelle (Visé), Liège Province, B.
M. Houssa.

SEM photographs of destinezite crystals.
Haut-le-Wastia (Anhée), Namur Province, B.
Royal Belgian Institute of Natural Sciences.

Originating from a small village in southern Belgium, **Pierre Joseph Destinez**, a man of modest birth and without any university degree, started his professional career as a laboratory worker to the department of mineralogy and geology of the University of Liège (ULg) (late 1872).

On May 30, 1876, he promoted to *conservateur préparateur* of the mineralogical and paleontological collections of the same department. Endowed with strong observation powers, manual skill, scientific turn of mind and aptitude for fieldwork, he undertook many field-trips, seizing the opportynity to collect many rock samples and fossils of the Palaeozoic and in particular specimens of the Carboniferous. In the end he became a real expert of the fauna and flora of the Carboniferous. Several fossils of the Palaeozoic of Belgium were even named after him.

The publication record of Destinez numbers tens of articles, some written in co-authorship with staff members of the department of geology of the ULg (Giuseppe Cesàro, Jules Fraipont, Maximillien Lohest, etc.). This not only reflects his great expertise of the palaeontology of the Carboniferous but also his ability to work successfully with his employers who eagerly took advantage of the interesting palaeontological material he collected during his lifetime to increase the knowledge of the scientific community. Hence, it is of little wonder, that in 1886 he was awarded by the *Société géologique de Belgique* (Liège) a unique Honorary Life Membership in recognition of his contribution to the study of the Carboniferous of Belgium and his dedication to this society.
Destinez died in Liège on April 7, 1911.

Biography

Lohest, M. (1911) *Annales de la Société géologique de Belgique*, **38**, B201-202.

DEWALQUITE -> **ARDENNITE-(As)**

Gustave Dewalque (1828-1905).

Salmchâteau (Vielsalm), Luxembourg Province, Belgium.

IMA: D

France: Musée de Minéralogie, Ecole des Mines, Paris - Type: *dewalquite* [donated by Pisani (with label). Pisani obtained this specimen from the Adam collection (N°2241) in 1873].

Pisani, F. (1872) Sur un nouveau silico-aluminate de manganèse vanadifère trouvé à Salm-Château, en Belgique. *Comptes rendus de l'Académie des Sciences, Paris,* **75**, 1542-1544.
Pisani, F. (1873) Analyse de la dewalquite de Salm-Château, en Belgique. *Comptes rendus de l'Académie des Sciences, Paris,* **77**, 329-333.

We refer to ardennite-(As) for the dispute between Félix Pisani and Arnold von Lasaulx about the discovery of ardennite/*dewalquite.*

Gilles Joseph **Gustave Dewalque** was born in Stavelot, in the Liège Province, on December 2, 1828. From 1844 onward he studied sciences and medicine at the University of Liège (ULg). In 1851 he was appointed *préparateur* to Frédéric-Antoine Spring, who held the chair of human and comparative physiology. One year later, he graduated as a doctor in medicine and in 1854 earned his doctor's degree in natural sciences. Dewalque then decided to continue his medical education in Paris but at this juncture a cholera epidemic broke out in Liège. Dewalque reconsidered his original idea to leave for France and decided to help the distressed inhabitants of Liège.

In 1855, Dewalque was appointed tutor to the course of mineralogy and geology given by André Dumont (*see*: dumontite) at the ULg and he also became curator of the university collections. After the sudden death of Dumont (early 1857), he acted as an interim professor. Six months later he received a special doctorate in mineral sciences. In October 1857, he was appointed extraordinary professor and took responsibility of the courses on mineralogy, geology and palaeontology. By the end of 1865, Dewalque became a full professor, a position he retained until his nominal retirement in 1897. He was succeeded by Max Lohest (*see*: *lohestite*). Dewalque passed away at the age of 80 on November 3, 1905, in Stavelot.

Dewalque's main research subject was the stratigraphy of the sedimentary formations of Belgium. The publication, in 1868, of a book entitled *Prodrome d'une description de la Belgique*, containing the explanatory texts of the geological map previously published by Dumont, was merely one of his merits. This book made him well-known at home and abroad. He was the promotor of the geological map of Belgium (scale 1:40,000) as well and founded the *Société géologique de Belgique* (1873). Apart from geology and palaeontology, Dewalque was also deeply interested in mineralogy, geophysics, hydrology, botany, meteorology, medicine and public health.

Biography

Liber memorialis. L'Université de Liège depuis sa fondation, col. 809-810, 1869. Liège: J.-G. Carmanne.

Lohest, M. (1911) *Annales de la Société géologique de Belgique*, **38**, B77-126.

Lohest, M. (1911) *Annuaire de l'Académie royale des Sciences, des Lettres et des Beaux-Arts de Belgique*, **77**, 53-103.

Liber memorialis. L'Université de Liège de 1867 à 1935, Notices biographiques, **II**, 31-36, 1936. Liège: Rectorat de l'Université.

Fourmarier, P. (1966) *Biographie nationale*, **33**, col. 255-261.

Bibliography

Liber memorialis. L'Université de Liège depuis sa fondation, col. 810-816, 1869. Liège: J.-G. Carmanne.

Lohest, M. (1911) *Annales de la Société géologique de Belgique*, **38,** B127-158.

Liber memorialis. L'Université de Liège de 1867 à 1935, Notices biographiques, **II,** 37-59, 1936. Liège: Rectorat de l'Université.

DRUGMANITE

Julien Drugman (1875-1950).

$Pb_2Fe^{3+}(PO_4)(PO_3OH)(OH)_2$

Monoclinic.

Richelle (Visé), Liège Province, Belgium.

Belgium:
Laboratory of Mineralogy, University of Liège - Type: HT: 19347.
Royal Belgian Institute of Natural Sciences, Brussels - Type: CT RN 5210.

IMA N°: 78-081

Van Tassel, R., Fransolet, A.-M. & Abraham, K. (1979) Drugmanite, $Pb_2(Fe^{3+},Al)(PO_4)_2(OH).H_2O$, a new mineral from Richelle, Belgium. *Mineralogical Magazine.* **43**, 463-467.

http://www.handbookofmineralogy.org/pdfs/drugmanite.pdf

Sheaf of drugmanite crystals with chisel-like terminations.
Richelle (Visé), Liège Province, B.
Royal Belgian Institute of Natural Sciences.
Cotype (N° N5210).

Aggregate of tabular drugmanite crystals.
Richelle (Visé), Liège Province, B.
Laboratory of Mineralogy, University of Liège.
Holotype (N° 19347).

Aggregates of tabular drugmanite crystals.
Richelle (Visé), Liège Province, B.
Royal Belgian Institute of Natural Sciences. Cotype (N° N5210).

Julien Drugman was born in Brussels on May 28, 1875. He was educated in the East Midlands of England (Derbyshire). Subsequently, he studied chemistry at the University of Bonn, where he received his doctor's degree in 1898. On his return to England, he became Master of Science at the University of Manchester (1906). He then served as a lecturer in chemistry at Huddersfield Technical College, West Yorkshire (1906-1907), and as a research chemist at the Institution of Gas Engineers, University of Leeds (1907).

Despite his solid knowledge of chemistry, Drugman gradually became more and more attracted by mineralogy and decided to carry out research in this speciality at the Oxford Mineralogical Laboratory under Herbert Lister Bowman. It was here that he published his first paper on twin crystals (1911). To get thoroughly conversant with mineralogy and crystallography, Drugman returned to Germany, where in 1912-1913 he worked under Victor Goldschmidt in Heidelberg, and with Paul Heinrich von Groth in Munich. In the meantime (1910), he joined the Royal Belgian Museum of Natural History in Brussels (since 1948 named Royal Belgian Institute of Natural Sciences, RBINS), where he accepted a position as a scientific associate to the Department of Mineralogy. He kept this post until the end of his professional career.

Drugman, who died on July 1, 1950, in Brussels, had sufficient means to travel extensively. During his numerous travels in Europe and overseas, both professionally and privately, he collected about 4,000 mineral specimens, a collection containing many superb and outstanding samples. As a mineral collector, his main goal was to gather as many different crystal forms and twins as possible. By Drugman's generous bequest, his rich and crystallographically most interesting collection, as well as all documents belonging to it, was passed on to the RBINS at his death, where it is known as the '*Drugman Collection*'. It also includes specimens deriving from the mineral collection of Anthyme Daimeries, former professor of analytical chemistry and crystallography at the Free University of Brussels.

Biography

Denaeyer, M.-E. (1951) *Bulletin de la Société française de Minéralogie et de Crystallographie*, **74**, 1-7.

Bibliography

Denaeyer, M.-E. (1951) *Bulletin de la Société française de Minéralogie et de Crystallographie,* **74**, 8-9.

FERRISTRUNZITE

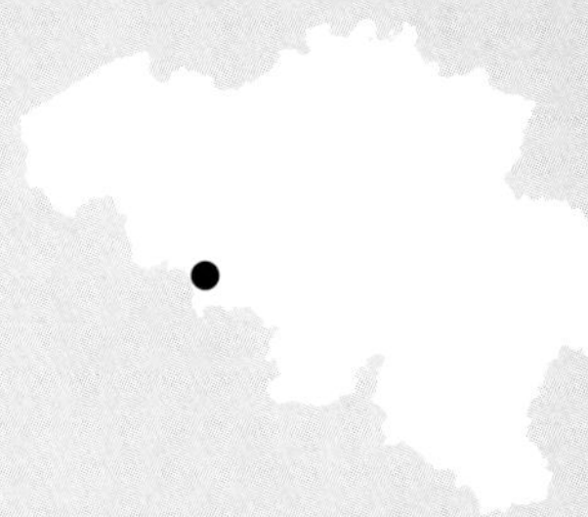

Ferric iron in its composition and relation to strunzite.
Strunzite refers to Hugo Strunz (1910-2006), a German mineralogist famous for his classification of minerals (Mineralogische Tabellen).

$Fe^{3+}(Fe^{3+})_2(PO_4)_2(OH)_3.5H_2O$

Triclinic.

Mont des Groseilliers, Blaton (Bernissart), Hainaut Province, Belgium.

USA: National Museum of Natural History, Washington - Type: HT N° 162499.

IMA N°: 86-023

Van Tassel, R. (1966) Minéraux secondaires phosphatés ferrifères (strunzite, beraunite, strengite, phosphosidérite, cacoxénite) de Blaton, Hainaut. *Bulletin de la Société belge de Géologie, de Paléontologie et d'Hydrologie*, **75**, 38-48.
Peacor, D.R., Dunn, P.J., Simmons, W.B. & Ramik, R.A. (1987) Ferristrunzite, a new member of the strunzite group, from Blaton, Belgium. *Neues Jahrbuch für Mineralogie, Monatshefte,* **10**, 453-457.

http://www.handbookofmineralogy.org/pdfs/ferristrunzite.pdf

Initially described as strunzite (Van Tassel, 1966), the investigations of Peacor *et al.* (1987) led to the discovery of a new mineral: ferristrunzite, isostructural with ferrostrunzite and strunzite.

Ferristrunzite is one of the most abundant phosphate minerals in the rich phosphate paragenesis on the southern slope of the Nimy-Péronnes canal. These phosphate minerals also occur as secondary minerals on Visean-Namurean shales and phtanites of the Mont des Groseilliers in Blaton.

SEM photograph of radiating acicular ferristrunzite crystals.
Blaton (Bernissart), Hainaut Province, B.
Royal Belgian Institute of Natural Sciences.

Aggregates of radiating acicular ferristrunzite.
Blaton (Bernissart), Hainaut Province, B.
J. Bourdeaudhui.

Fibrous pale beige ferristrunzite crystals.
Blaton (Bernissart), Hainaut Province, B.
L. Schelfhout.

Small star-like aggregates of acicular ferristrunzite on black shale.
Blaton (Bernissart), Hainaut Province, B.
Royal Belgian Institute of Natural Sciences (N° 4771).

Hollow prismatic ferristrunzite.
Blaton (Bernissart), Hainaut Province, B.
J. Bourdeaudhui.

Detail of hollow prismatic ferristrunzite.
Blaton (Bernissart), Hainaut Province, B.
J. Bourdeaudhui.

Fan-shaped aggregate of acicular beige ferristrunzite.
Blaton (Bernissart), Hainaut Province, B.
L. Schelfhout.

FRAIPONTITE

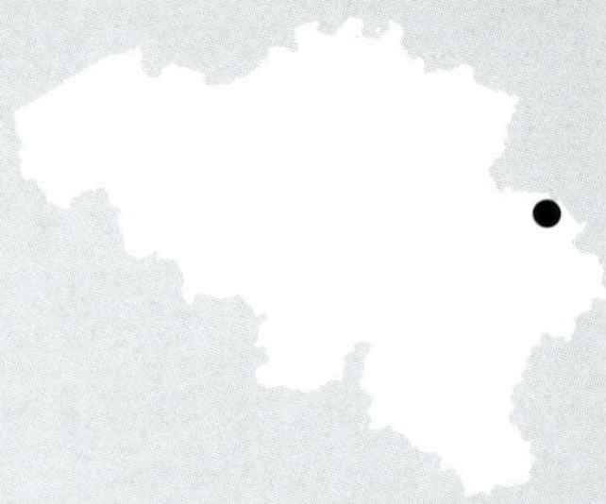

Julien Fraipont (1857-1910) and Charles de Fraipont (1883-1946).

$(Zn,Al)_3(Si,Al)_2O_5(OH)_4$

Monoclinic.

(?) Altenberg/Vieille-Montagne, Kelmis/La Calamine, Liège Province, Belgium.

Belgium: Laboratory of Mineralogy, University of Liège - Type: HT: 13727.

IMA: G

Cesàro, G. (1927) Sur la fraipontite, silicate basique hydraté de zinc et d'aluminium. *Annales de la Société géologique de Belgique,* **50**, 106-110.

Fransolet, A.-M. & Bourguignon, P. (1975) Données nouvelles sur la fraipontite de Moresnet (Belgique). *Bulletin de la Société française de Minéralogie et de Cristallographie*, **98**, 235-244.

http://www.handbookofmineralogy.org/pdfs/fraipontite.pdf

The origin of this mineral remains dubious. Since the original description, no other specimens of fraipontite have been found near Moresnet. It is possible that the specimens were found near Moresnet but originated from an unknown site abroad. Indeed, when local mining of zinc ores ceassed at this locality, the refining industry imported ores from mines all over the world. It is even likely that fraipontite was found among these imported ores (Michel Deliens, personal communication).

Banded layers of brownish fraipontite in bluish smithsonite. Kelmis/La Calamine, Liège Province, B. Royal Belgian Institute of Natural Sciences (N° L19092).

Julien Fraipont was a native of Liège where he was born on August 17, 1857. At the completion of his secondary education and in spite of a great dedication to natural sciences he started working with his father, who had founded and was the director of the *Crédit Général Liégeois*. Concurrently, Julien enrolled at the University of Liège (ULg) where he took the courses on zoology given by Edouard Van Beneden. Upon his advice, Julien soon retired from financial business and studied natural sciences at the university. He became a doctor in natural sciences in 1878. When that study was completed he subsequently became student-assistant, *préparateur* (1878) and assistant (1881) under Van Beneden.

When, in 1884, Gustave Dewalque (*see*: *dewalquite*) was relieved of the course of palaeontology, Fraipont accepted to teach this subject. Hardly one year later, he also assumed responsibility for teaching zoogeography and systematic zoology. In 1886 he was appointed an extraordinary professor and three years later full professor. His academic career was at its peak when he was elected rector of the ULg in 1909. Unfortunately, he deceased a few months later in Liège (March 22, 1910).

The scientific work of Julien Fraipont focused on systematics and morphology of different kinds of animals and from 1883 onwards also on palaeontological topics in relation to the Devonian and Carboniferous of Belgium. In parallel with these research subjects, he was also interested in prehistory and archaeology. At the request of his colleague Max Lohest (*see*: *lohestite*) and the prehistorian Marcel De Puydt, he also agreed to study the Neanderthal skulls that both scientists discovered and excavated in the middle of the 1880s in a cave at Spy (Jemeppe-sur-Sambre), in the Namur Province.

Biography

Fourmarier, P. (1919) *Annales de la Société géologique de Belgique*, **41,** B337-344.
Lohest, M., Julin, Ch. & Rutot, A. (1925) *Annuaire de l'Académie royale des Sciences, des Lettres et des Beaux-Arts de Belgique,* **91**, 131-190.
Liber memorialis. L'Université de Liège de 1867 à 1935, Notices biographiques, **II**, 186-190, 1936. Liège: Rectorat de l'Université.
Ubaghs, G. (1973-1974) *Biographie nationale*, **38,** col. 221-224.

Bibliography

Fourmarier, P. (1919) *Annales de la Société géologique de Belgique*, **41,** B345-350.
Lohest, M., Julin, Ch. & Rutot, A. (1925) *Annuaire de l'Académie royale des Sciences, des Lettres et des Beaux-Arts de Belgique,* **91**, 191-197.
Liber memorialis. L'Université de Liège de 1867 à 1935, Notices biographiques, **II**, 190-193, 1936. Liège: Rectorat de l'Université.

Born on March 21, 1883 at Chênée (Liège), **Charles de Fraipont,** son of Julien Fraipont (see above), received his secondary education at the Jesuit College in Liège. In 1908 he graduated as a mining engineer at the *Faculté Technique* of the University of Liège (ULg). In 1909 he was appointed assistant for palaeontology and in 1911 became curator of the palaeontological collections of the ULg. Shortly before the start of World War I, he received the degree of special doctor in palaeontological sciences at the Faculty of Sciences of the Liège alma mater. At the close of the war, Charles de Fraipont took charge of the courses on animal palaeontology and plant palaeontology at the Faculty of Sciences, where he succeeded Paul Cerfontaine and Alfred Gilkinet. Concurrently he became *suppléant* of geology and mineralogy at the *Institut Agronomique de l'Etat* in Gembloux. In 1921 he was appointed extraordinary professor and in 1923 full professor at the Faculty of Sciences of the ULg.

Charles de Fraipont was the founder of the *Ecole libre d'anthropologie* attached to the ULg, where he was responsible for palaeoanthropology, palaeoethnography, physical anthropology, human palaeontology and anthropology. When the changeover to Dutch was considered at Ghent University, Charles de Fraipont became dedicated to '*Gand-français*'. In 1931, being elected senator of the Liège district, he gave up the greater part of his academic duties. He passed away in Brussels on April 15, 1946.

Charles de Fraipont had a lifelong interest in prehistory. This is hardly surprising as his father Julien was closely involved in the discovery and study of the so-called '*man of Spy*'. Charles' study of a child's skull, found by Philippe-Charles Schmerling, a Belgian doctor and anatomist, at the turn of 1829 and 1830 in a cave at Engis (halfway between Huy and Liège), was one of his most outstanding scientific contributions. He demonstrated that the fossil bones he examined did not belong not to a modern human but to a Neanderthal.

Biography

Liber memorialis. L'Université de Liège de 1867 à 1935, Notices biographiques, **II**, 303-305, 1936. Liège: Rectorat de l'Université.

Liber memorialis. L'Université de Liège de 1936 à 1966, Notices historiques et biographiques, **II,** 413-416, 1967. Liège: Rectorat de l'Université.

Bibliography

Liber memorialis. L'Université de Liège de 1867 à 1935, Notices biographiques, **II**, 305-310, 1936. Liège: Rectorat de l'Université.

Liber memorialis. L'Université de Liège de 1936 à 1966, Notices historiques et biographiques, **II,** 417, 1967. Liège: Rectorat de l'Université.

Franquenies (Court-Saint-Etienne), Walloon Brabant Province, Belgium.

IMA: D

Anthoine, R. & P. (1942) Les assises de Mousty et de Villers-la-Ville du bassin supérieur de la Dyle. *Annales de la Société géologique de Belgique,* **66**, M53-180.

Van Tassel, R. (1944) Notes minéralogiques. I. Occurrence de minéraux sulfatés récents sur des schistes de Mousty, à Franquenies. *Bulletin du Musée royal d'Histoire naturelle de Belgique,* **20**, 1-23.

Van Tassel, R. (1949) Notes minéralogiques. IV. L'identité entre slavíkite et franquenite. *Bulletin du Musée royal d'Histoire naturelle de Belgique,* **25**, N° 7, 1-16.

The mineral was discovered in a quarry where Cambrian schists are exposed.
Initially, *franquenite* was erroneously identified as a mixture of sulphur and gypsum (Anthoine, 1942). Van Tassel (1949) proved that franquenite is identical to slavíkite. It is closely associated with other sulphate minerals including melanterite, pickeringite, ferropickeringite and gypsum.
It is likely that the mineral crystallized at Franquenies as a result of neoformation.

GALAPEKTITE -> MONTMORILLONITE

The etymology remains uncertain. Many scientists believe that *galapektite* derives from the Greek terms *gala*, meaning milk, and *piktos/pektos,* referring to its congealed appearance.

Angleur (Liège), Liège Province, Belgium.

IMA: D

Breithaupt , A. (1832) Vollständige Characteristik des Mineral-Systems: dritte sehr bereicherte Ausgabe: 307 p., Dresden.

Ross, C.S. & Kerr, P.F. (1934) Halloysite and allophane. *United States Geological Survey, Professional Papers,* **185 G**, 135-148.

Faust, G.T. (1957) A study of the montmorillonite variety galapektite. *Journal of the Washington Academy of Sciences*, **47**, N° 5, 143-146.

Fleischer, M. (1957) New mineral names. *The American Mineralogist*, **42**, 919-921.

Next to halloysite Breithaupt (1832) also reported the results of his investigations on a new mineral he named *galapektite*. Both minerals came from Angleur.
Investigations and analyses of Ross & Kerr (1934) proved that *galapektite* was in fact montmorillonite. The authors nevertheless mentioned an incorrect origin for the specimen: *Anglar, France*.
The conclusions of Ross & Kerr (1934) were confirmed by investigations done by Faust (1957), resulting in a proposal formulated by Fleischer (1957) to discredit *galapektite*.

GRAULICHITE-(Ce)

Jean-Marie Graulich (1920-2001).

$CeFe^{3+}_3(AsO_4)_2(OH)_6$

Trigonal.

Hourt quarry, Grand-Halleux (Vielsalm), Luxembourg Province, Belgium.

Belgium:
Laboratory of Mineralogy, University of Liège - Type: HT N° 20325.
Royal Belgian Institute of Natural Sciences, Brussels - Type: CT N° RN 6419.

IMA N°: 2002-001

Hatert , F., Lefèvre, P., Pasero, M. & Fransolet, A.-M. (2002) Graulichite-(Ce), a new arsenate mineral from the Stavelot Massif, Belgium. *European Journal of Mineralogy,* **15**, 733-739.

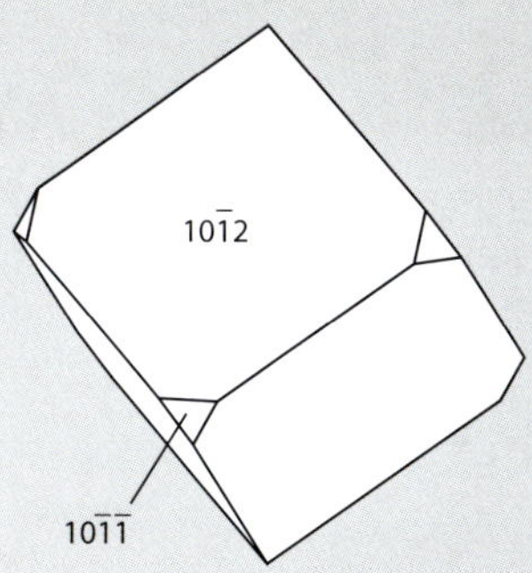

after Hatert *et al.* (2002)

Aggregates of graulichite-(Ce) on quartz.
Hourt quarry, Grand-Halleux (Vielsalm),
Luxembourg Province, B.
Laboratory of Mineralogy, University of Liège.

*Aggregates of small brown yellow graulichite-(Ce) in quartz cavity.
Hourt quarry, Grand-Halleux (Vielsalm), Luxembourg Province, B.
M. Blondieau*

*Aggregates of rhombohedral graulichite-(Ce) on quartz.
Hourt quarry, Grand-Halleux (Vielsalm), Luxembourg Province, B.
M. Blondieau.*

*Aggregates of small greenish grey graulichite-(Ce) on quartz.
Hourt quarry, Grand-Halleux (Vielsalm), Luxembourg Province, B.
M. Blondieau.*

Jean-Marie Graulich was born in Liège on May 4, 1920, and studied at the university in his birthplace. He obtained degrees in mining engineering and engineering geology in 1947 and 1948, respectively. He began his professional career as an *assistant volontaire* and an assistant to the geological institute of the University of Liège. In 1949, he became a pre-doctoral researcher to the Belgian National Fund for Scientific Research (NFWO-FNRS) in Brussels and two years later served as a consultant engineer geologist to the *Société minière de la Pertulosa* (Italy). In 1952, Graulich joined the Geological Survey of Belgium (GSB) in Brussels, where he was employed until his mandatory retirement in 1985. He first served the GSB as a geologist with a temporary contract and, after several promotions, finally retired from this institution as chief geologist/director. Following his retirement he continued to work on problems he already dealt with during his professional career. Graulich passed away on March 17, 2001.

From 1949 until 1960, Graulich's travels took him to many countries, including Algeria, Greece, continental and insular Italy, and Sweden. Here, his fieldwork and research activities focused on a variety of geological problems related to mineralization processes, tectonics, stratigraphy, sedimentology, metamorphism, hydrogeology and geophysical prospection. The experience he gained during these scientific missions formed an excellent preparation for the tasks awaiting him during his employment at the GSB. These comprised subjects mainly relating to the geology of southern Belgium and dealing with almost every field of geological research. Being a student of Paul Fourmarier (*see*: fourmarierite), he had a lifelong interest in and a predilection for problems in relation to tectonics. Graulich's many fields of interest are manifested in the numerous reports, papers and maps he published during his professional life and after his formal retirement.

Biography

Delmer, A. & Da Silva, A.-C. (2003) *Geologica Belgica*, **6/3-4**, 73-76.

Bibliography

Delmer, A. & Da Silva, A.-C. (2003) *Geologica Belgica*, **6/3-4**, 76-80.

HALLOYSITE-10Å and HALLOYSITE-7Å

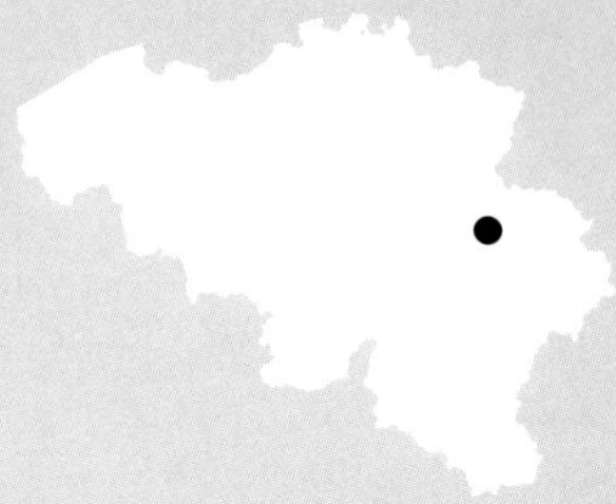

Jean-Baptiste d'Omalius d'Halloy (1783-1875).

Angleur (Liège), Liège Province, Belgium.

France: Muséum national d'Histoire naturelle, Paris - Type: T: 28133
(donated by d'Omalius d'Halloy).

HALLOYSITE-10Å (former *endellite*)

$Al_2Si_2O_5(OH)_4.2H_2O$

Monoclinic.

IMA: G

HALLOYSITE-7Å (former *metahalloysite*)

$Al_2Si_2O_5(OH)_4$

Monoclinic.

IMA: G

Berthier, P. (1826) Analyse de l'halloysite. *Annales de Chimie et de Physique*, Paris, **32**, 332-335.

Ross, C.S. & Kerr, P.F. (1934) Halloysite and allophane. *United States Geological Survey, Professional Papers,* **185 G**, 135-148.

Bates, T.F., Hildebrand, F.A. & Swineford, A. (1950) Morphology and structure of endellite and halloysite. *The American Mineralogist,* **35**, 463-484.

Faust, G.T. (1955) The endellite-halloysite nomenclature. *The American Mineralogist*, **40**, 1110-1118.

Burke, E.A.A.J. (2006) A mass discreditation of GQN minerals. *The Canadian Mineralogist*, **44**, 1557-1560.

http://www.handbookofmineralogy.org/pdfs/halloysite.pdf

Halloysite - a very important clay mineral - was discovered by d'Omalius d'Halloy. It was first described by Berthier (1826) who decided to name it after its discoverer.
For a long time, halloysite was confused with endellite, the hydrated form of halloysite.
Faust (1955) reviewed the nomenclature of both clay minerals. He endowed halloysite with the chemical formula $Al_2Si_2O_5(OH)_4$, and endellite with the chemical formula $Al_2Si_2O_5(OH)_4.2H_2O$.
A few years ago, the Commission on New Minerals, Nomenclature and Classification of the International Mineralogical Association [Burke (2006)] discredited endellite as mineral species as by then the name halloysite-10Å was in current use to designate the hydrated form of halloysite.

TEM photograph showing cylindric halloysite (associated with muscovite and goethite). Central Kongo Province, D. R. Congo. Laboratory of Mineralogy and Petrology, Ghent University.

Crust of bluish white halloysite. Welkenraedt, Liège Province, B. Royal Belgian Institute of Natural Sciences (N° 5455).

Crust of bluish white halloysite. Welkenraedt, Liège Province, B. Royal Belgian Institute of Natural Sciences (N° 5455).

As suggested by its latinized name, Jean-Bernard d'Omalius d'Halloy, father of Jean-Baptiste, had his roots in Omal (Geer), in the Hesbaye. Yet at least part of the year he lived at the patrimonial property of Halloy (Ciney), in the Condroz region, with his wife Sophie de Thier de Skeuvre. The rest of the year the family lived in Liège. It was there that **Jean-Baptiste** Julien **d'Omalius d'Halloy**, their only child, was born on February 16, 1783. He received a classical education at home but at the age of 18, being well-born, was sent to Paris to become thoroughly acquainted with daily life of upper-class people, with literature, arts, etiquette, eloquence, etc. Two other journeys to the French capital were to follow (1803, 1805). When he resided in Paris, Jean-Baptiste was a frequenter of the *Jardin des Plantes* and the *Muséum national d'Histoire naturelle,* where he attended lectures on chemistry and geology given by noted French scientists such as Alexandre Brongniart, Baron Georges Cuvier, Count Antoine de Fourcroy, René Haüy, Etienne de La Ville-sur-Illon (Count of Lacépède), Jean-Baptiste de Monet (Knight de Lamarck), and Nicolas Vauquelin.

In 1804 d'Omalius started with geological field observations throughout the French Empire. In the following decade, one can deduce from his field-notes, he managed to cover on foot and with the geological hammer near at hand about 25,000 km. This is equivalent to about thirty times the distance from Belgium to the Pyrenean Mountains. Till today the significance of his work can hardly be overestimated. His *Essai sur la Géologie du Nord de la France* went to press in 1808 and before the close of 1813 he succeeded in pre-paring a geological map of the French Empire. Unfortunately, it was not published until 1822. After Napoleon's overthrow, d'Omalius was appointed governor of the Namur Province and had to interupt all his favourite scientific activities. The revolution of 1830 marked the end of his governorship. He retreated to his rural estate at Halloy and began a new period of field studies and research with respect to geology, geography and ethnography. In 1848, he was elected member of the Belgian Senate and three years later became vice-president. D'Omalius d'Halloy, statesman and pioneer of modern geology in Belgium, passed away on January 15, 1875, in Brussels, at the advanced age of 92. His death was the consequence of a stroke he suffered about one year earlier while doing geological observations in the valley of the Senne River.

Biography

Dupont, E. (1876) *Annuaire de l'Académie royale des Sciences, des Lettres et des Beaux-Arts de Belgique*, **42**, 181-288.

Gosselet, J. (1878-1879) *Annales de la Société Géologique du Nord*, **6**, 457-477.

Anonymous (1885) d'Omalius d'Halloy. 84 p. Namur: L. Raikem.

Dupont, E. (1897) d'Omalius d'Halloy 1783-1875, 1-86. Musée royal d'Histoire naturelle de Belgique. Bruxelles: Polleunis et Ceuterick.

Guequier, J. (1901) *Biographie nationale*, **16**, col. 157-166.

Fourmarier, P. (1967) Florilège des Sciences en Belgique pendant le XIXe siècle et le début du XXe, 430-438. Bruxelles: Académie royale de Belgique, Classe des Sciences.

Bibliography

Dupont, E. (1876) *Annuaire de l'Académie royale des Sciences, des Lettres et des Beaux-Arts de Belgique*, **42**, 288-295.

Dupont, E. (1897) d'Omalius d'Halloy 1783-1875. 87-95. Musée royal d'Histoire naturelle de Belgique. Bruxelles: Polleunis et Ceuterick.

HOPEITE

Thomas Charles Hope (1766-1844), former professor of chemistry at Edinburgh, Scotland.

$Zn_3(PO_4)_2.4H_2O$

Orthorhombic.

Altenberg/Vieille-Montagne, Kelmis/La Calamine, Liège Province, Belgium.

France: Muséum national d'Histoire naturelle, Paris - Type: CT: 104.452 (Des Cloizeaux collection, 1879).

IMA: G

Brewster, D. (1822) New mineral from Altenberg, near Aachen. *Edinburgh Philosophical Journal,* **6**, 185-186.

Brewster, D. (1824) Description of hopeite, a new mineral from Altenberg, near Aix-la-Chapelle. *Transactions of the Royal Society of Edinburgh,* **10**, 107-111.

Lévy, A. (1843) Description de plusieurs espèces minérales appartenant à la famille du zinc. *Annales des Mines, Paris*, 4ième Série, **4**, 517.

Buttgenbach, H. (1918) Contribution à l'étude des minéraux belges. *Annales de la Société géologique de Belgique*, **42**, M93-124.

Goldschmidt, V. (1918) Atlas der Krystallformen. **IV**, 212 p., 133 Tafeln, Heidelberg: Carl Winters Universitätsbuchhandlung.

http://www.handbookofmineralogy.org/pdfs/hopeite.pdf

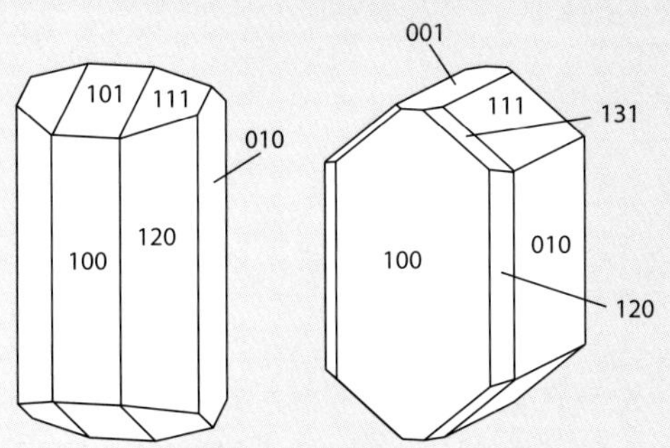

after Brewster (1824)

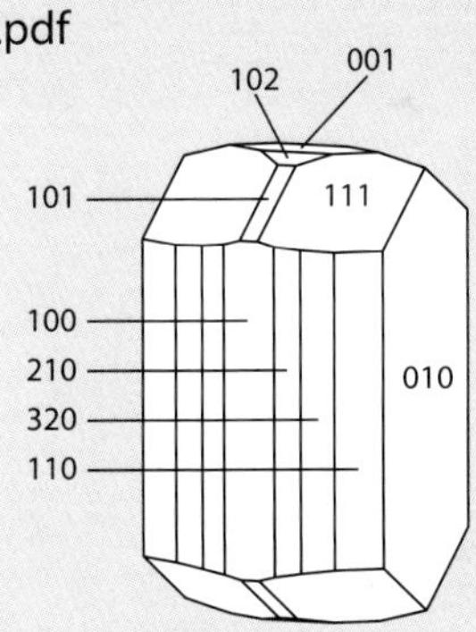

after Lévy (1843)

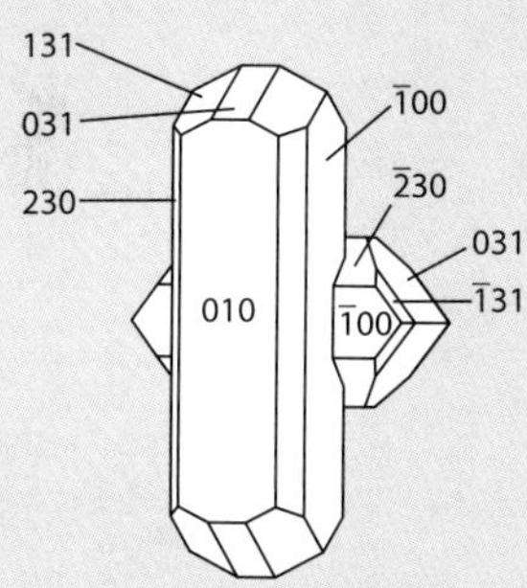

after Buttgenbach (1918)

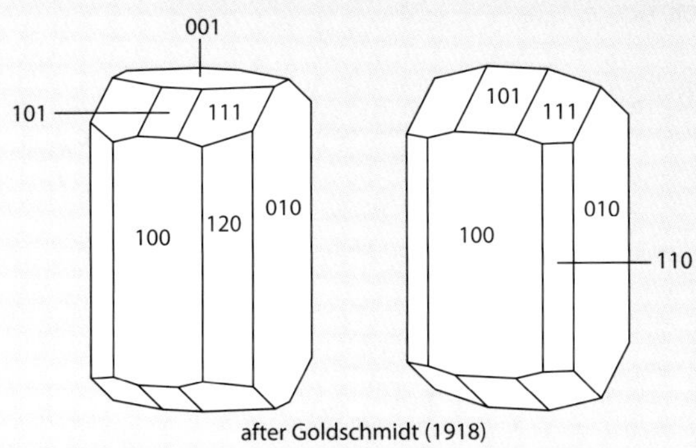

after Goldschmidt (1918)

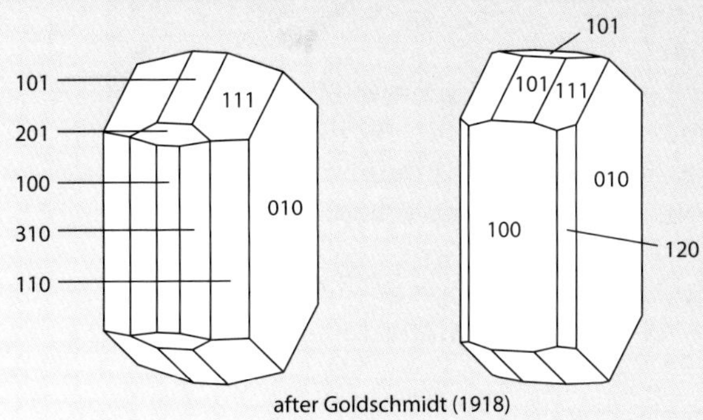

after Goldschmidt (1918)

Brewster (1822) proved that the mineral he received from the British mineral dealer John Henry Heuland, labelled as stilbite, was not stilbite at all. He also found out that the French mineralogist René-Just Haüy considered the same mineral to be a variety of stilbite to which he gave the name *stilbite duovigésimale*.
In 1824 Brewster described it as a new mineral species and named it hopeite. It was found a few years before at Altenberg /Vieille-Montagne, Kelmis/La Calamine.

Chisel-like prismatic hopeite crystal.
Kelmis/La Calamine, Liège Province, B.
Royal Belgian Institute of Natural Sciences (N° 2988).

Prismatic hopeite crystals.
Kelmis/La Calamine, Liège Province, B.
Royal Belgian Institute of Natural Sciences (N° 6499).

Double terminated prismatic hopeite crystal.
Kelmis/La Calamine, Liège Province, B.
Royal Belgian Institute of Natural Sciences (N° 6499).

Remark:
In 1907 Leonard James Spencer of the Mineral Department of the British Museum in London described for the first time parahopeite, a triclinic dimorph of hopeite. It came from Broken Hill mines in North Rhodesia (modern Zambia). Parahopeite was thus not discovered in Belgium, neither honours a Belgian citizen, but we feel free to mention it here, as a complement to the description of hopeite.

Spencer, L. (1908) On hopeite and other zinc phosphates and associated minerals from the Broken Hill mines, North-Western Rhodesia. *Mineralogical Magazine*, **15**, 1-38.

KLEMENTITE -> ferrian CHAMOSITE

Constantin Klement (1856-1902).

Vielsalm, Luxembourg Province, Belgium.

IMA: D

Tschermak, G. (1890-1891) Die Chloritgruppe. *Sitzungsberichte der Kaiserlichen Akademie der Wissenschaften in Wien. Mathematisch-Naturwissenschaftliche Klasse.* **99**, Abt. 1, 174-266, and **100**, Abt. 1, 29-107.

Mélon, J. (1938) Description des chlorites et clintonites belges. *Mémoires de l'Académie royale de Belgique, Classe des Sciences*, **17**, 5-73.

Hey, M.H. (1954) A new review of the chlorites. *Mineralogical Magazine*, **30**, 277-292.

Bayliss, P. (1975) Nomenclature of the trioctahedral chlorites. *The Canadian Mineralogist,* **13**, 178-180.

Tschermak (1890-1891) gave the name *klementite* to leptochlorites from Vielsalm and Salmchâteau, analysed by Klement. Mélon (1938) also used this name to describe specimens from Vielsalm and Regné.
Hey (1954) classified *klementite* as a variety of *thuringite.*
Thuringite was discredited being identical with chamosite (Bayliss, 1975).

Constantin Klement was born on June 27, 1856, at Karlsberg (Moravia), in the former Austro-Hungarian Empire. He studied chemistry in Vienna under Ernst Ludwig, and mineralogy under Gustav Tschermak. After earning his doctor's degree in sciences, he obtained a position as an assistant to the laboratory of the K.k. *Chemisch-Physiologische Versuchsstation* at Klosterneuburg, near Vienna. As a result of his close contacts with Ludwig and Tschermak, and thanks to several official journeys in the Austro-Hungarian capital in the 1870s and 1880s, Alphonse Renard (*see*: renardite) was well informed about the skills and meticulous care of Klement with regard to quantitative chemical analysis. This may have encouraged Renard to ask Edouard Dupont - then director of what is now the Royal Belgian Institute of Natural Sciences - to invite Klement to come and work at the institute in Brussels and to assist him in his research at the department of mineralogy and lithology.

Klement accepted Renard's proposal, left his native country and was employed as an *aide-naturaliste* on March 1, 1882. In hardly six weeks, Klement set up the chemical laboratory of Renard's department and on April 13 of the same year he started his quantitative analytical work. The very first sample he analysed was a piece of *bastonite* (*see:* this work) from Libramont. Klement was also the first scientist to carry out a chemical analysis of the meteorite that fell at Sint-Denijs-Westrem, near Ghent, on June 7, 1855. According to the notes found in his laboratory book, he also did chemical analyses of deep-sea sediments dredged during the famous *Challenger* expedition (*see*: renardite).

When Renard was appointed full professsor at Ghent University (1888), Klement became responsible for running the laboratory in Brussels, previously headed by Renard. By September 1889 he became curator to the department of mineralogy and lithology, a position he held until his death in Etterbeek (Brussels) on July 31, 1902. Born an Austro-Hungarian citizen, Klement obtained the Belgian nationality five years before he passed away.

Bibliography

Lohest, M. (1911) Discours prononcé aux funérailles de M. Pierre Destinez. *Annales de la Société géologique de Belgique*, **38**, B201-202

KONINCKITE

Laurent-Guillaume de Koninck (1809-1887).

$Fe^{3+}PO_4.3H_2O$

Tetragonal.

Richelle (Visé), Liège Province, Belgium.

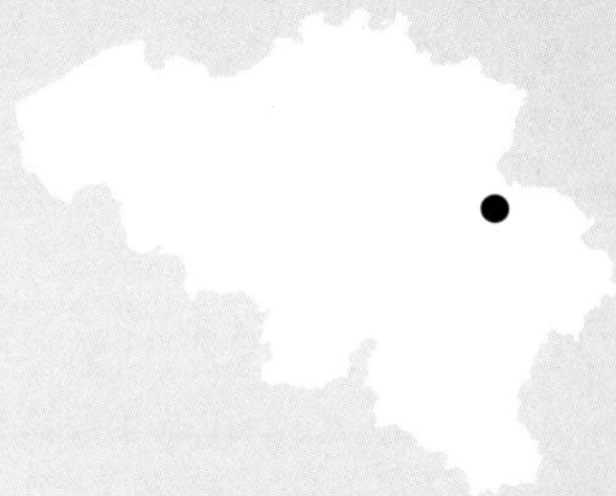

France: Muséum national d'Histoire naturelle, Paris - Type: T: 125157 & 135475 (donated by Cesàro).
Belgium: Laboratory of Mineralogy, University of Liège (HT/CT N° 9251-9257).

IMA: G

Cesàro, G. (1883) Mémoire traitant de la koninckite, nouveau phosphate ferrique hydraté, de la formule de la richellite et de l'oxyfluorure de fer. *Annales de la Société géologique de Belgique,* **11**, 247-257.

http://www.handbookofmineralogy.org/pdfs/koninckite.pdf

SEM photograph of a globular aggregate of koninckite.
Richelle (Visé), Liège Province, B.
Royal Belgian Institute of Natural Sciences.

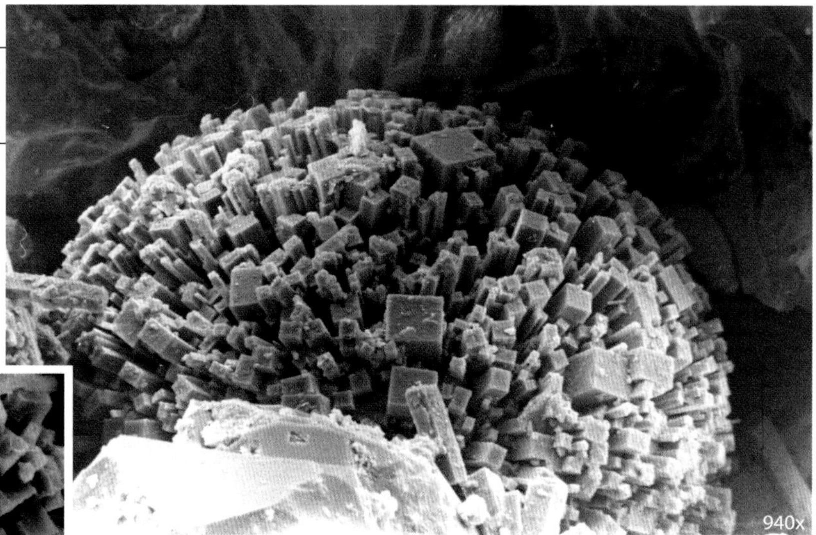

SEM photograph of aggregates of koninckite.
Richelle (Visé), Liège Province, B.
Laboratory of Mineralogy, University of Liège.

5 mm

Botryoidal koninckite.
Richelle (Visé), Liège Province, B.
Royal Belgian Institute of Natural Sciences (N° 5065).

8 mm

Brownish globular koninckite.
Richelle (Visé), Liège Province, B.
Laboratory of Mineralogy, University of Liège.

Clear brownish globular koninckite.
Richelle (Visé) , Liège Province, B.
M. Houssa.

4 mm

Laurent-Guillaume de Koninck was born in Leuven on May 3, 1809. He received his early education in his native town and then enrolled at the local university. At the age of 22 he was 'candidate' in mathematical and physical sciences, and doctor in medicine, in 'the art of childbirths' and in pharmacy. In 1831 he became *préparateur* at the Catholic University of Leuven (CUL), where he worked in the department headed by Jean-Baptiste Van Mons, a chemist of international renown. A few years later, he travelled to France and Germany to improve his knowledge of chemistry. In 1835 he worked in Paris under Louis Joseph Gay-Lussac and Baron Louis Jacques Thénard, and doing so gained deeper insight into systematic and experimental anorganic chemistry. In Berlin, he attended lectures on physical chemistry given by Eilhard Mitscherlich, whereas in Giessen he worked under Baron Justus von Liebig on problems regarding organic chemistry. De Koninck also visited first-rate laboratories and outstanding scientists in Bonn, Marburg and Heidelberg.

Upon his return to Belgium, de Koninck was appointed *agrégé* at Ghent University (December 1835), where he lectured industrial chemistry. In October 1836, he joined the University of Liège (ULg), where over the years he became responsible for the courses

on general chemistry, chemistry applied to arts, and also lectured an in-depth course on organic chemistry for students preparing to become doctor in natural sciences. In 1838, de Koninck was appointed extraordinary professor and in 1856 became a full professor. In 1846 and 1847, de Koninck was asked to teach an optional course on palaeontology but later on he decided to give up this part of his teaching programme to avoid the doctrinal controversy he had with André Dumont (*see*: dumontite) being pushed to extremes. Dumont was in doubt about the scientific value of palaeontological observations for age determination in geology and believed that the physical and mineralogical properties of rocks gave much more certainty than information provided by the study of the fossils they contained. De Koninck, on the other hand, was confident that fossils were of major importance to geologists and offered more prospects. The continuing heated scientific dispute between upholders of and opponents to Dumont's theory was finally settled in de Koninck's favour.

After retirement from active teaching and research at the ULg in 1876, de Koninck continued his palaeontological studies and spent his remaining time and efforts in writing down the results of the scientific work that had occupied his interests over the past 50 years until the close of his life. He passed away in Liège on July 15, 1887.

Ever since 1843, de Koninck's research centered on problems related to animal palaeontology, in particular on the study of Palaeozoic faunas from all over the world. He paid particularly attention to the fossils of Carboniferous age of Belgium and for this period alone he identified over 700 new species. As a token of honour, several palaeontologists worldwide named a wide variety of newly discovered plant and animal species after him.

Biography

Liber memorialis. L'Université de Liège depuis sa fondation, col.788-790, 1869. Liège: J.-G. Carmanne.
Fraipont, J. (1887) *Annales de la Société géologique de Belgique,* **14,** B189-248.
Oehlert, D. (1887-1888) *Bulletin de la Société géologique de France,* 3me série, **16**, 466-474.
Dupont, E. (1891) *Annuaire de l'Académie royale des Sciences, des Lettres et des Beaux-Arts de Belgique*, **57**, 437-475.
Dupont, E. (1891) Notice sur Laurent-Guillaume De Koninck. 1-39. Bruxelles: F. Hayez.
Liber memorialis. L'Université de Liège de 1867 à 1935, Notices biographiques, **II**, 9-14, 1936. Liège: Rectorat de l'Université.
Van Luppen, J. (1981) *Nationaal biografisch Woordenboek*, **9**, col. 431-436.

Bibliography

Liber memorialis. L'Université de Liège depuis sa fondation, col.790-796, 1869. Liège: J.-G. Carmanne.
Fraipont, J. (1887) *Annales de la Société géologique de Belgique,* **14,** B249-255.
Oehlert, D. (1887-1888) *Bulletin de la Société géologique de France,* 3me série, **16**, 474-477.
Dupont, E. (1891) *Annuaire de l'Académie royale des Sciences, des Lettres et des Beaux-Arts de Belgique,* **57**, 476-483.
Dupont, E. (1891) Notice sur Laurent-Guillaume De Koninck. 40-47. Bruxelles: F. Hayez.
Liber memorialis. L'Université de Liège de 1867 à 1935, Notices biographiques, **II**, 14-18, 1936. Liège: Rectorat de l'Université.

LOHESTITE -> a mixture of KAOLINITE, PYROPHYLLITE, SERICITE and CHLORITE

Maximilien (Max) Lohest (1857-1926).

Several places in the Salm valley, Luxembourg Province, Belgium.

IMA: D

Anten, J. (1923) Le Salmien métamorphique du sud du massif de Stavelot. *Mémoires (in-4°) de l'Académie royale de Belgique*, Classe des Sciences, 2e Série, **5**, 34 p. + map.

Theunissen, K. (1970) L'andalousite et ses phases de transformation dans la région de Vielsalm. *Annales de la Société géologique de Belgique*, **93**, 363.

An amorphous material surrounding andalusite grains and occurring in metamorphic Salmian (Lower Ordovician) rocks was designated as *lohestite* by Anten (1923).
More recent analyses (Theunissen, 1970) have shown that *lohestite* is a mixture of kaolinite, pyrophyllite, sericite and chlorite.

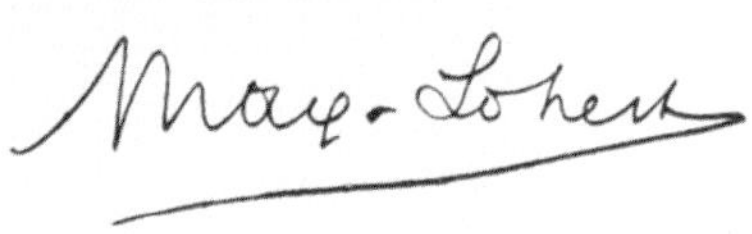

Maximilien (Max) Lohest was born in Liège on September 8, 1857. At the completion of his secondary education at the Jesuit College of his birthplace, he enrolled for one year at the Faculty of Philosophy and Literature (Arts) of the University of Liège (ULg) to study philosophy. In 1883, he graduated as a honorary engineer of mines at the Technical Faculty of the ULg and the following year he was appointed assistant to Gustave Dewalque (*see*: *dewalquite*). In 1890, Lohest was asked to teach a voluntary course on combustibles and phosphates and this course was transformed in 1893 into a non-compulsory course on applied geology and hydrology. It was the first step in his academic career.

At the retirement of Dewalque (1897), Lohest became head of the general geology section and director of the School of Mines. From 1900 onward, Lohest and some of his colleagues, including Julien Fraipont (*see*: fraipontite) and Giuseppe Cesàro (*see*: cesàrolite), did their utmost best to convince the Belgian government to organize engineering geology studies at the ULg. In the end, they achieved their goal. The creation of the School of Anthropology of Liège, shortly after the end of World War I, can also be attributed to Maximilien.

At his demand Lohest was relieved of the course on applied geology and hydrology in 1920, in order to focus his scientific work exclusively on problems relating to general geology and on a number of subjects he was particularly interested in. Yet, in those days his health already deteriorated and, a few months before retiring from the university, he passed away on December 6, 1926 in Liège.

The scientific turn-over of Lohest was impressive, encompassing all fields of geological sciences, as he published papers on mineralogy, petrography, physical geography, palaeontology, prehistoric anthropology, along with many others on general geology, applied geology and hydrology. He was also strongly involved, either as author or as co-author, in the realization of several parts of the geological map of Belgium at scale 1:40,000. Although Lohest travelled a lot, he never did any geological fieldwork in the former Belgian Congo. This did however not prevent him to publish ten papers and reports on different aspects of the geology of the former Belgian colony.

Biography

Prost *et al.* (1926-1927) Discours prononcés aux funérailles de M. Max Lohest, ancien président. *Annales de la Société géologique de Belgique*, **50**, B57-B84.
Liber memorialis. L'Université de Liège de 1867 à 1935, Notices biographiques, **II**, 206-216, 1936. Liège: Rectorat de l'Université.
Fourmarier, P. (1953) *Annuaire de l'Académie royale de Belgique*, **CXIX**, 279-343.
Fourmarier, P. (1955) *Biographie coloniale belge,* **IV**, col. 535-536.
Delmer, A. (2001) *Nouvelle biographie nationale,* **6,** 281-282.

Bibliography

Liber memorialis. L'Université de Liège de 1867 à 1935, Notices biographiques, **II**, 217-234, 1936. Liège: Rectorat de l'Université.
Fourmarier, P. (1953) *Annuaire de l'Académie royale de Belgique* **CXIX**, 344-386.
Fourmarier, P. (1955) *Biographie coloniale belge* **IV**, col. 536-537 (only papers with regard to the former Belgian Congo).

MORESNETITE -> a mixture of HEMIMORPHITE and SAUCONITE

Moresnet, Liège Province, Belgium.

Altenberg/Vieille-Montagne, Kelmis/La Calamine, Liège Province, Belgium.

IMA: D

Risse, H. (1865) Über den Moresnetit, ein neues Zinkoxyd-Tonerdesilikat von Altenberg bei Aachen. *Naturhistorischer Verein der preussischen Rheinlande und Westphalens, Correspondenblatt,* Bonn, **22**, 98-99.

Kenngott, G.A. (1868) Übersichte der Resultate mineralogische Forschungen (1852-1868) für die Jahren 1862-65, Leipzig 1868.

Ross, C.S. (1946) Sauconite - A clay mineral of the montmorillonite group. *The American Mineralogist*, **31**, 411-424.

Faust, G.T. (1951) Thermal analysis and X-ray studies of sauconite and some zinc minerals in the same paragenetic association. *The American Mineralogist*, **36**, 795-822.

Following the first description of *moresnetite* by Risse (1865), Kenngott (1868) claimed that *moresnetite* was a mixture of hemimorphite and clay.
Microscopic research by Ross (1946) showed that *moresnetite* was a mixture of hemimorphite, micas and chlorite. Differential thermal analysis performed by Faust (1951) indicated that the mixture may also include a substantial amount of sauconite.

For a concise history of Moresnet, the zinc ore mining and its minerals, *see: belgite.*

OTTRÉLITE

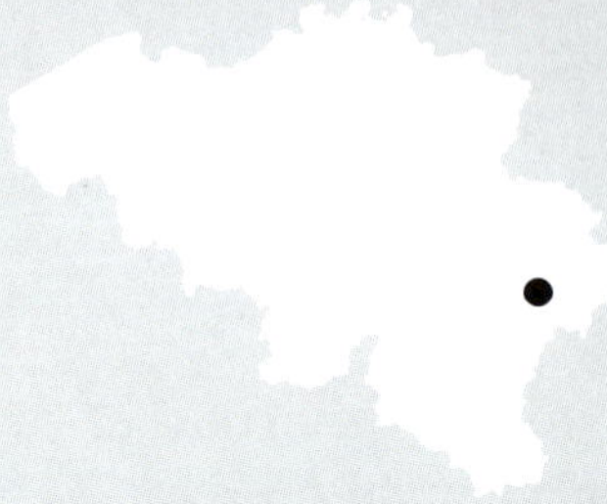

Ottré (Vielsalm), Luxembourg Province, Belgium

$(Mn^{2+})Al_2O(SiO_4)(OH)_2$

Monoclinic (single-crystal investigation shows a mixture of monoclinic and triclinic polymorphs).

Ottré (Vielsalm), Luxembourg Province, Belgium.

France: Muséum national d'Histoire naturelle, Paris - Type: T: 104479 (collection of Alfred Des Cloizeaux).

IMA: G

Dethier, L. (1818) Le guide des curieux qui visitent les Eaux de Spa. Liège, 1819, 14p.
Damour, M. & Des Cloizeaux, A. (1842) De l'ottrélite, nouvelle espèce minérale. *Annales des Mines, Paris, 4ième série,* **2**, 357.
Mélon, J. (1938) Description des chlorites et clintonites belges. *Mémoires de l'Académie royale de Belgique, Classe des Sciences,* **17**, 5-73.
Fransolet, A.-M. (1978) Données nouvelles sur l'ottrélite d'Ottré, Belgique. *Bulletin de Minéralogie,* **101**, 548-557.

http://www.handbookofmineralogy.org/pdfs/ottrelite.pdf

This mineral, discovered by Dethier in 1809 and described for the first time by Damour & Des Cloizeaux (1842), has a long and complex history due to a lack of reliable analyses. Names such as *diallage métalloide, ottrélite, salmite* or *impure manganese chloritoid* appear in literature.
According to Mélon (1938) ottrélite is a valid mineral species, chemically similar to a manganese chloritoid but optically very different from the latter.
The history of the naming and the opposite ideas about the nature of ottrélite have been outlined by Fransolet (1978). His investigations proved that ottrélite is a valid species and represents the manganese end-member of the chloritoid group.

Thin section of an ottrélite-bearing rock (transmitted plane-polarized light); note the hourglass structure. Ottré (Vielsalm), Luxembourg Province, B. Laboratory of Mineralogy and Petrology, Ghent University.

SEM photograph of ottrélite crystals.
Ottré (Vielsalm), Luxembourg Province, B.
Laboratory of Mineralogy, University of Liège.

Aggregates of ottrélite.
Ottré (Vielsalm), Luxembourg Province, B.
Laboratory of Mineralogy, University of Liège (N° 19277).

Ottrélite aggregate on quartz.
Ottré (Vielsalm), Luxembourg Province, B.
M. Fouassin.

Detail of a deep green ottrélite crystal.
Ottré (Vielsalm), Luxembourg Province, B.
University of Liège (N° 19277).

PUMPELLYITE-(Al)

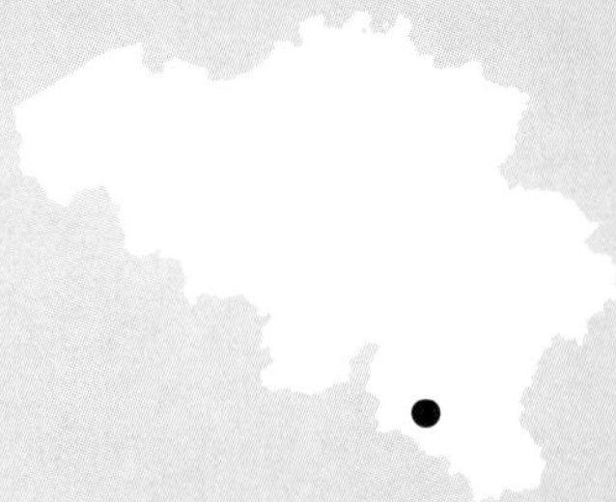

Al-dominant analogue of pumpellyite.
Honours Raphael Pumpelly (1837-1923), an American geologist and explorer.

$Ca_2Al_3(SiO_4)(Si_2O_7)(OH,O)_2.H_2O$

Monoclinic.

Flèche quarry, Bertrix, Luxembourg Province, Belgium.

Belgium: Laboratory of Mineralogy, University of Liège Type: HT N°20327.

IMA N°: 2005-016

Hatert, F., Pasero, M., Perchiazzi, N. & Theye, Th. (2007) Pumpellyite-(Al), a new mineral from Bertrix, Belgian Ardennes. *European Journal of Mineralogy*, **19**, 247-253.

Aggregates of radiating greyish blue pumpellyite-(Al).
Bertrix (Vielsalm), Luxembourg Province, B.
M. Blondieau.

Aggregates of radiating
greyish blue pumpellyite-(Al).
Bertrix (Vielsalm), Luxembourg Province, B.
M. Houssa.

RICHELLITE

Richelle (Visé), Liège Province, Belgium.

$Ca(Fe^{3+})_2(PO_4)_2(OH,F)_2$

Amorphous.

Richelle (Visé), Liège Province, Belgium.

Belgium: Laboratory of Mineralogy, University of Liège - Type: HT: 9361 and 9375.
France: Muséum national d'Histoire naturelle, Paris - Type: HT: 125234 (donated by Cesàro as part of HT).

IMA: Q

Cesàro, G. & Despret, G. (1883): La richellite. Nouvelle espèce minérale des environs de Visé. *Annales de la Société géologique de Belgique,* **10**, 36-44.
McConnell, D. (1963) Thermocrystallization of richellite to produce a lazulite structure (calcium lipscombite). *The American Mineralogist*, **48**, 300-307.

http://www.handbookofmineralogy.org/pdfs/richellite.pdf

Richellite was first discovered in a cavity affecting a breccious Dinantian limestone (Cesàro & Despret, 1883). According to the original description, the mineral forms a kind of cement and occurs in association with halloysite and allophane. Thanks to later studies of the Richelle site, gypsum and koninckite were added to the paragenesis. The habit of richellite is diverse: as a coating covering the rock, or it is botryoidal, nodular, coralloidal or layered, often with a golden coating. McConnell (1963) noted that some analytical techniques cannot be applied to study it and, when available, chemical and diffraction data lack precision. Because of the great variation in the chemical composition and the amorphous nature of the analysed specimens, the International Mineralogical Association considers richellite as a questionable mineral. To get a clear-cut status, the examined specimens require further examination.

Crust of brown richellite.
Richelle (Visé), Liège Province, B.
H. Daniels.

Small globular richellite.
Richelle (Visé), Liège Province, B.
P. du Ry.

Red brown richellite cementing rock fragments.
Richelle (Visé), Liège Province, B.
Laboratory of Mineralogy, University of Liège (N° 9366).

Detail of botryoidal richellite.
Richelle (Visé), Liège Province, B.
M. Houssa.

Thin layers of richellite.
Richelle (Visé), Liège Province, B.
Laboratory of Mineralogy, University of Liège (N° 9252).

SALMITE -> manganiferous CHLORITOID

Salm River, Luxembourg Province, Belgium.

IMA: D

Prost, E. (1884) Sur la salmite de Dumont, chloritoïde manganésifère. *Annales de la Société géologique de Belgique,* **11**, 93-97.

De Rauw, H. (1912) Note sur la salmite, le rutile et la tourmaline d'Ottré. *Annales de la Société géologique de Belgique*, **38**, B209-214.

Mélon, J. (1938) Description des chlorites et clintonites belges. *Mémoires de l'Académie royale de Belgique, Classe Sciences*, **17**, 5-73.

Theye, T. & Fransolet, A.-M. (1994) Virtually pure ottrélite from the region of Ottré, Belgium. *European Journal of Mineralogy*, **6**, 547-555.

Prost (1884) was the first to analyse mineral specimens collected by de Koninck on dumps at Vielsalm. He concluded that the examined mineral was a member of the chloritoid group. In the mineral collection of the University of Liège he found the same mineral, collected years before by André Dumont and labelled *salmite*, without any reference to literature about this mineral.

De Rauw (1912) reported analyses of *salmite* from Ottré and drew attention on its high manganese content. Mélon (1938) described *salmite* as a manganiferous chloritoid.

Recent investigations (Theye & Fransolet, 1994) did not restrain *salmite* as a valid mineral species, but as a variety of chloritoid.

STAVELOTITE-(La)

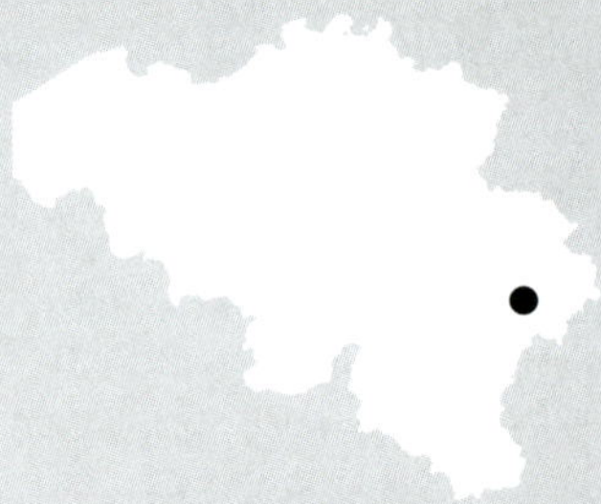

Refers to the geological entity called "Stavelot Massif" in the Belgian Ardennes. Stavelot is a town in the Liége Province, Belgium.

$La_3(Mn^{2+})_3Cu^{2+}(Mn^{3+},Fe^{3+}, Mn^{4+})_{26}(Si_2O_7)_6O_{30}$

Trigonal.

Le Coreux, Salmchâteau (Vielsalm), Luxembourg Province, Belgium.

Germany: Institut für Geologie, Mineralogie und Geophysik, Ruhr University, Bochum - Type: HT N°25169.

IMA N°: 2004-014

Bernhardt, H-J., Armbruster, Th., Fransolet, A.-M. & Schreyer, W. (2005) Stavelotite-(La), a new lanthanum-manganese-sorosilicate mineral from the Stavelot Massif, Belgium. *European Journal of Mineralogy*, **17**, 703-714.

Stavelotite-(La) is a mineral with a very complex crystal structure. It was found in a late-stage quartz vein affecting manganiferous phyllites of Ordovician age and cropping out in the southeastern part of the Stavelot Massif. So far this mineral was exclusively recognized in thin sections.

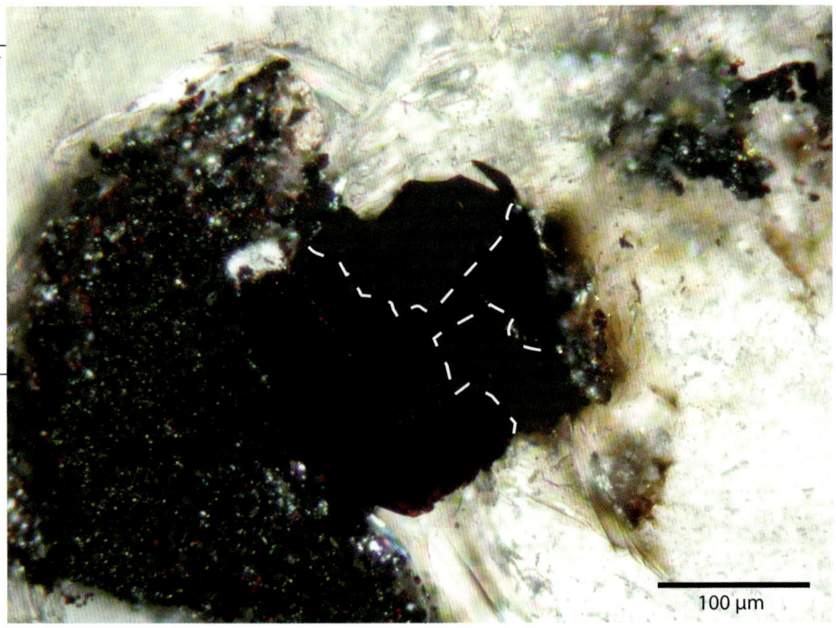

Thin section - reflected light photograph of an irregular mass consisting of at least two individual crystals of stavelotite-(La). Grain boundaries are marked with white dashes. Le Coreux, Salmchâteau (Vielsalm), Luxembourg Province, B. Institut für Geologie, Mineralogie und Geophysik, Ruhr University, Bochum, Germany.

Photograph courtesy of Heinz-Jürgen Bernhardt, Ruhr University, Bochum, Germany.

René Van Tassel (°1916).

$Al_4(PO_4)_3(OH)_3.9H_2O$

Orthorhombic.

Bihain (Vielsalm), Luxembourg Province, Belgium.

Belgium: University of Liège - Type: HT.

IMA N°: 86-016

Fransolet, A.-M. (1987) La vantasselite, $Al_4(PO_4)_3(OH)_3.9H_2O$, une nouvelle espèce minérale du Massif de Stavelot, Belgique. *Bulletin Minéralogique,* **110**, 647-656.

http://www.handbookofmineralogy.org/pdfs/vantasselite.pdf

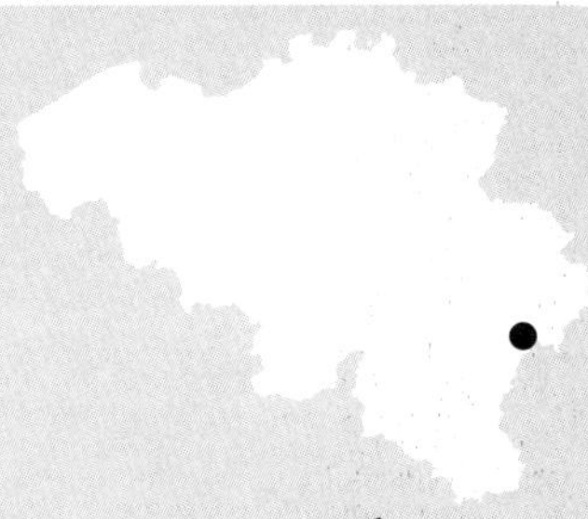

after Fransolet (1987)

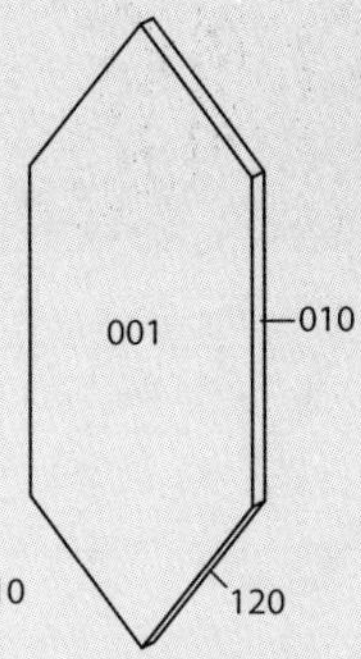

after Fransolet (1987)
(rotated position)

Before it was recognized as a new mineral species, vantasselite was known as *mica carré* (square mica) .

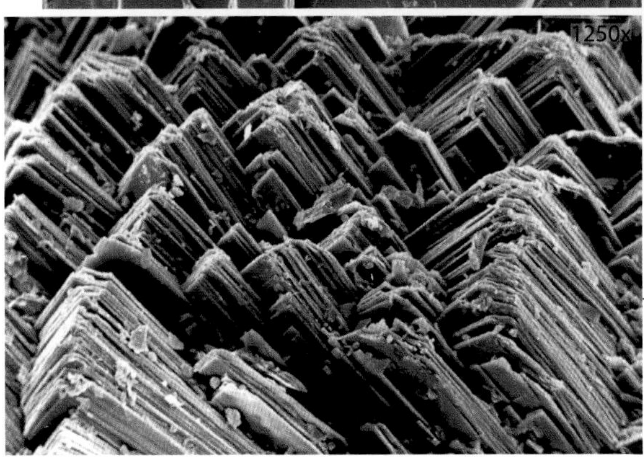

SEM photographs of vantasselite aggregates.
Bihain (Vielsalm), Luxembourg Province, B.
Royal Belgian Institute of Natural Sciences.

Cross-section through spherical aggregates of vantasselite.
Bihain (Vielsalm), Luxembourg Province, B.
E. Van Der Meersche.

Cross-section through spherical aggregates of vantasselite.
Bihain (Vielsalm), Luxembourg Province, B.
P. Rondelez.

Intact aggregate of vantasselite crystals.
Bihain (Vielsalm), Luxembourg Province, B.
M. Blondieau.

Aggregates of vantasselite.
Bihain (Vielsalm), Luxembourg Province, B.
J. Detaille.

René Van Tassel was born on May 11, 1916 in Antwerp. After his early schooling he started university studies at the Catholic University of Leuven (CUL). He earned his doctor's degree in sciences (chemistry group) in 1939. In 1939-1940, to broaden his knowledge of mineralogy, he worked as a voluntary student in the laboratory headed by Jacques Thoreau (*see:* thoreaulite) at the Leuven alma mater. One year later, he accepted a position at the Royal Belgian Museum of Natural History (nowadays called the Royal Belgian Institute of Natural Sciences, RBINS) in Brussels. He would spend the rest of his professional career at the department of mineralogy and petrography of this institute, at first as an *aide-naturaliste* and finally as director and head of department. He kept this position until his official retirement in 1981.

In 1949-50, Van Tassel held a British Council scholarship. He took this opportunity to do research and to attend lectures at the department of mineralogy and petrology of the University of Cambridge, U.K. During this sojourn, he described a new mineral that he discovered in ejected blocks of peralkaline granite from Ascension Island, in the Atlantic. To Van Tassel, it was self-evident to name this zirconium silicate new to science dalyite in honour of Reginald Aldworth Daly. In the past this professor emeritus of Harvard University had strongly contributed to unravel the geological history and petrology of Ascension Island. In 1971, Van Tassel was appointed extraordinary professor at the Faculty of Sciences of the *Katholieke Universiteit Leuven* (KUL), where he was in charge of a course on geochemistry until he retired at the RBINS. After officially retiring, he remained scientifically active and continued to publish in both scientific and popularizing journals.

The scientific contribution of Van Tassel to the mineralogy of Belgium is simply impressive. Apart from dalyite, he was also the first to describe calclacite and drugmanite. By proposing the name drugmanite for a mineral he found at Richelle, Belgium, Van Tassel expressed his sympathy and gratitude to Julien Drugman (*see*: drugmanite). This former associate of the mineralogy section at the RBINS was during his lifetime particularly interested in twinning problems and bequeathed his unique mineral collection to the museum in Brussels. Van Tassel was also the first to recognize β-FeOOH and Mn-free strunzite as new mineral species, years later, these two minerals were named and fully described by other researches.

VIAENEITE

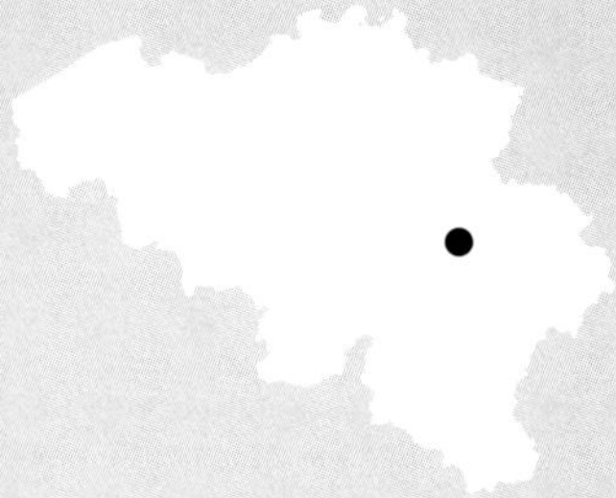

Willy Viaene (1940-2000).

$(Fe,Pb)_4S_8O$

Monoclinic.

La Mallieue, Engis, Liège Province, Belgium.

Belgium:
Royal Belgian Institute of Natural Sciences, Brussels - Type: HT: RN 6380.
Katholieke Universiteit Leuven (KUL) - Type: CT: KUR32, ONB2, ONB3.
Poland: Faculty of Geology, Academy of Mining & Metallurgy, Kraków - Type: CT: EMP9.

IMA N°: 93-051

Kucha, H., Osuch, W. & Elsen, J. (1996) Viaeneite, $(Fe,Pb)_4S_8O$, a new mineral with mixed sulphur valencies from Engis, Belgium. *European Journal of Mineralogy,* **8**, 93-102.
Dillen, H. (1996) Viaeneiet, nieuw... uit België, *Geonieuws*, **21**, 177-185.

http://www.handbookofmineralogy.org/pdfs/viaeneite.pdf.

Viaeneite was discovered during a microscopic study of polished sections of ore samples from a zinc-lead deposit in Engis, a municipality about 20 km SW of Liège.
One of the characteristics of the sulphide minerals of this deposit are the different valencies of sulphur, due to changing reduction/oxidation circumstances during the ore formation.
Viaeneite presents small oval-shaped intergrown aggregates inside pyrite, pyrite-melnikovite, marcasite and sphalerite. Exposed to the air, it tarnishes and disintegrates within a few weeks.
In the paragenesis Kucha *et al.* (1996) discovered, beside the most abundant minerals sphalerite, pyrite, melnikovite, marcasite and galenite, also smaller amounts of goethite, zincian siderite, smithsonite, greigite and other Zn-Pb compounds with double sulphur valencies.

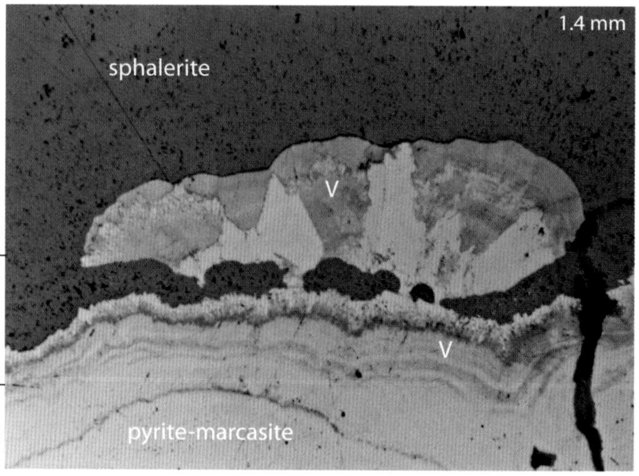

*Polished section : V = Viaeneite.
La Mallieue, Engis, Liège Province, B.
Katholieke Universiteit Leuven (KUL).
Photo courtesy of H. Dillen (1996).*

The following photographs show details of a polished ore sample from Engis, used for investigation and stored as holotype in the collection of the Royal Belgian Institute of Natural Sciences.

In the pictures one observes banded ochre layers of sphalerite, pyrite/marcasite as fan-shaped aggregates, and galenite as bluish black patches. As viaeneite is instable and forms very small inclusions in pyrite, marcasite, melnikovite and sphalerite, it is impossible to distinguish it on the photographs.

Born on January 22, 1940, in Staden, **Willy Viaene**'s youth was spent in West-Flanders. After his classical education, he enrolled at the Faculty of Sciences of the Catholic University of Leuven (CUL). Here, he received his licentiate's degree in Earth sciences (1963), and earned his PhD in geology (1967) under the guidance of Jules Moreau (*see*: moreauite), with a dissertation on the synthesis of sulphide minerals. After finishing his doctorate, he obtained a post-doctoral research mandate of the Belgian National Fund for Scientific Research in Brussels. Between 1968 to 1970 Willy Viaene performed post-doctoral studies and research under Gerhard Christian Amstutz at the Institute of Mineralogy and Petrology, University of Heidelberg (Germany), and under Gunnar Kullerud, at the Geophysical Laboratory in Washington, USA. These studies were funded by research fellowships awarded by the Alexander von Humboldt Foundation and the Carnegie Institution of Washington. In 1971, he was appointed visiting assistant professor to the Geoscience Department of Purdue University, West Lafayette, Indiana, USA. This allowed him to continue working in the USA for another eight months.

Back in Belgium, Viaene advanced rapidly through the academic environment of Leuven. He was promoted to lecturer in 1971 and four years later became professor. From 1979 onward, he served the *Katholieke Universiteit Leuven* (KUL) as a full professor. Up till his death he excelled in this function. Although his teaching duties at this institution mainly comprised the field of pure and applied mineralogy, he also taught the first-year course on general geology. After a very productive professional career and a lingering illness, he died on March 13, 2000, at the early age of 60.

Next to his work on ore deposits, sediment petro-graphy, diagenetic processes and industrial mineralogy (especially focusing on the study of clays, mortars and ceramic materials), Viaene also performed interdisciplinary research yielding an extensive if not unique knowledge and expertise in experimental mineralogy, geochemisty of carbonate rocks, pollution processes and archaeometry. He played a prominent part in several international projects sponsored by the European Economic Community. For many years, and on top of his major administrative responsibilities, he also actively participated in the interdisciplinary archaeological project performed by members of the KUL at the site of Sagalassos (Turkey).

Biography

Anonymous (2000) *Klei/Glas/Keramiek (Kgk),* Tijdschrift voor klei, glas en keramiek, **21**/4, 5.

Degryse P. & Sintubin, M. (2002) Contributions to the Geology of Belgium and Northwest Europe. Proceedings of the 1st Geologica Belgica International Meeting. *Aardkundige Mededelingen*, **12**, vi.

VISÉITE -> mixture of CRANDALLITE and other phases

Visé, Liège Province, Belgium.

IMA: D

Mélon, J. (1942) La viséite, nouvelle espèce minérale. *Annales de la Société géologique de Belgique,* **66**, B53-56.

Di Renzo, F. & Gabelica, Z. (1995) New data on the structure and composition of the silicoaluminophosphate viséite and a discreditation of its status as a zeolite. In: *Natural Zeolites '93: Occurrence, Properties, Use.* D. W. Ming and F. A. Mumpton, International Committee on Natural Zeolites, Brockport, New York, 173-185.

Kim, Y. & Kirkpatrick, R.J. (1996) Application of MAS NMR spectroscopy to poorly crystalline minerals: viseite. *Mineralogical Magazine,* **60**, 957-962.

Coombs, D. S., Alberti, A., Armbruster, T., Artioli, G., Colella, C., Galli, E., Grice, J.D., Liebau, F., Mandarino, J.A., Minato, H., Nickel, E.H., Passaglia, E., Peacor, D.R., Quartieri, S., Rinaldi, R., Ross, M., Sheppard, R.A., Tillmanns, E. & Vezzalini, G. (1997) Recommended nomenclature for zeolite minerals: Report of the subcommittee on zeolites of the International Mineralogical Association, Commission on New Minerals and Mineral Names, *The Canadian Mineralogist,* **35**, 1571-1606.

http://www.handbookofmineralogy.org/pdfs/viseite.pdf

Although Mélon (1942) described *viséite* as a new mineral species, it was already discovered by Giuseppe Cesàro in 1888.
Di Renzo & Gabelica (1995) pointed out that *viséite* was not a zeolite, and Kim & Kirkpatrick (1996) clearly demonstrated that the material they analysed was a mixture of crandallite and other phases, including opal.
Coombs *et al.* (1997) proved that viséite was neither a zeolite nor a mineral s.s.

WILLEMITE

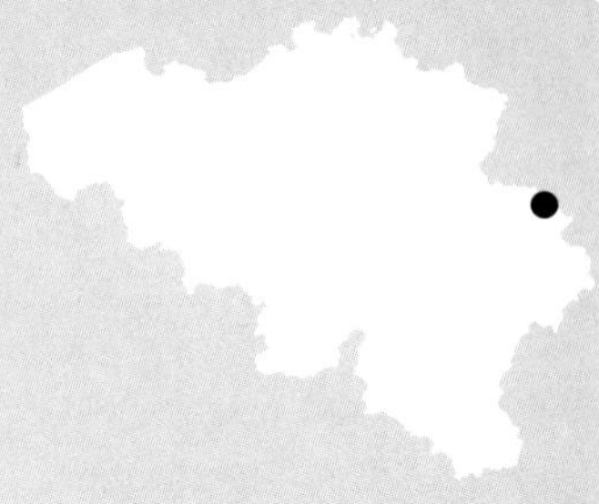

William I (1772-1843), King of the Netherlands (then also comprising the region that became the Kindom of Belgium in 1830).

Zn_2SiO_4

Trigonal.

Franklin-Sterling Hill, Ogdenburg, Franklin Mining District, Sussex County, New Jersey, USA.

Altenberg/Vieille-Montagne, Kelmis/La Calamine, Liège Province, Belgium.

France: Musée national d'Histoire naturelle, Paris - Type: T: 33346 & 104980 (donated by Lévy).

IMA: G

Vanuxem, L. & Keating, W. H. (1824) Observations upon some of the minerals discovered at Franklin, Sussex County, N.J.. *Journal of the Academy of Natural Sciences of Philadelphia*, **4**, 3-11.
Lévy, A. (1830) Lettre à l'éditeur. *Neues Jahrbuch für Mineralogie*, **1**, 71.
Lévy, A. (1843) Description de plusieurs espèces minérales appartenant à la famille du zinc. *Annales des Mines*, 4me *Série*, 507-520.
Cesàro, G. (1887) Notes sur quelques minéraux. *Annales de la Société géologique de Belgique*, **14**, CXLII-CXLIII.
Buttgenbach, H. (1947) Les minéraux de Belgique et du Congo Belge. 573 p., Paris: Dunod.

http://www.handbookofmineralogy.org/pdfs/willemite.pdf.

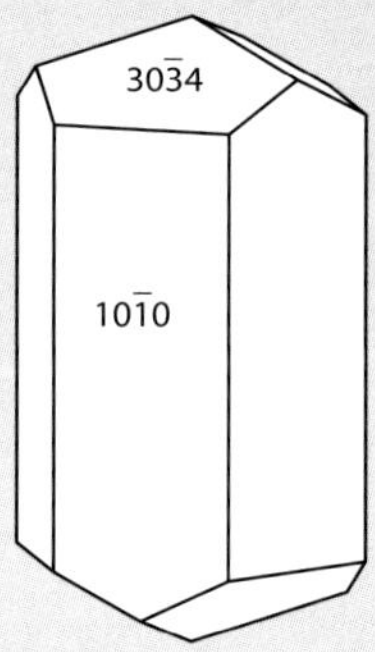

after Buttgenbach (1947)

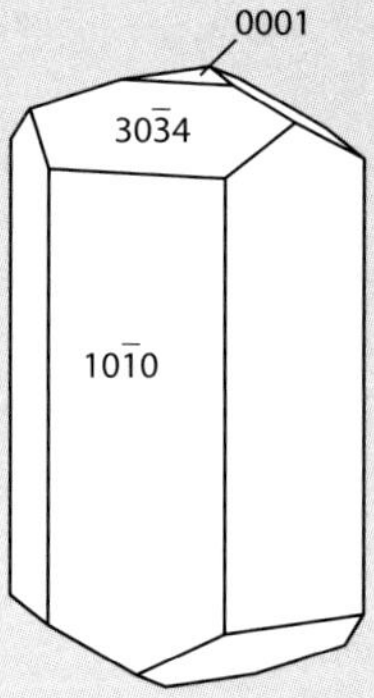

after Cesàro (1887)

The name of this zinc silicate of fairly rare occurrence was first used by the French mineralogist Armand Lévy (1830), who found it at the end of 1829, probably near Moresnet (*see: moresnetite*).
After a while, however, it became clear that the new mineral had already been recognized in former days as a distinct mineral species from New Jersey (Vanuxem & Keating, 1824), yet at that time it remained unnamed and was simply labelled as *siliceous oxide of zinc*.
The name willemite was suggested by Lévy hardly a few months before Belgium officially proclaimed its independence (November 1830). William I continued governing the southern part of The Netherlands and Lévy, who described several new mineral species, found it only fair to honour the king of his host country with the name of a mineral species.
Panebianco introduced the name *belgite* as replacement for willemite (*see: belgite*).

Transparent, double terminated willemite crystal.
Kelmis/La Calamine, Liège Province, B.
Laboratory of Mineralogy, University of Liège (N° 12918).

Aggregates of short-prismatic willemite with rust-coloured coating.
Kelmis/La Calamine, Liège Province, B.
Royal Belgian Institute of Natural Sciences (N° N3871).

Aggregate of transparent, prismatic willemite crystals with striated faces.
Kelmis/La Calamine, Liège Province, B.
Royal Belgian Institute of Natural Sciences (N° A5002).

SEM photographs of willemite crystals.
Kelmis/La Calamine, Liège Province, B.
Royal Belgian Institute of Natural Sciences.

PART II. MINERALS DISCOVERED ABROAD, NAMED AFTER BELGIANS

ANDRÉMEYERITE

André Meyer (1920-1965).

$Ba(Fe^{2+})_2Si_2O_7$

Monoclinic (pseudo-orthorhombic).

Nyiragongo volcano, Nord-Kivu Province, Democratic Republic of the Congo.

IMA N°: 72-005

Finland: Geological Museum of the University of Helsinki. HT: missing (private communication O. Tapani Rämö).

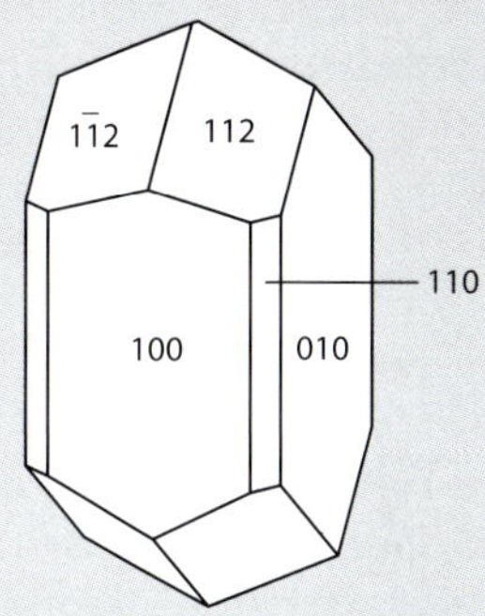

after Sahama *et al.* (1973)

Sahama, Th. G., Siivola, J. & Rehtijärvi, P. (1973) Andremeyerite, a new barium iron silicate from Nyiragongo, Zaire. *Bulletin of the Geological Society of Finland*, **45**, 1-8.
Cannillo, E., Mazzi, F. & Rossi, G. (1988) Crystal structure of andremeyerite, $BaFe(Fe,Mn,Mg)Si_2O_7$. *American Mineralogist*, **73**, 608-612.

http://www.handbookofmineralogy.org/pdfs/andremeyerite.pdf

During one of their expeditions in the fifties of last century, André Meyer and Thure Sahama collected blocks of a melilite-leucite-nephelinite on the southern rim of the Nyiragongo volcano. An unknown mineral, mainly forming irregular and very small crystals, exceptionally euhedral prismatic crystals in vesicles, was found in some of these blocks. Sahama investigated this new mineral and named it andrémeyerite in honour of his friend who died at an early age (Sahama *et al.*, 1973). Cannillo *et al.* (1988) examined the crystal structure of andrémeyerite using material from the holotype provided by Thure Sahama. This could explain the absence of any holotype material in the present-day collection in Helsinki.

Aggregate of pale green andrémeyerite crystals. Nyiragongo, Nord-Kivu Province, D.R. Congo. Royal Belgian Institute of Natural Sciences (Vanacker collection).

André Meyer was born on October 13, 1920 in Åbo (present-day Turku), Finland. He graduated from the Catholic University of Leuven in 1943 with a mining engineer degree and at the same institution he completed the engineering geology programme in 1944. The year 1945 marked the beginning of Meyer's professional career, when he accepted to work in Kivu (former Belgian Congo) with the *Société Belge de Recherches Minières en Afrique* (REMINA) and the *Comité National du Kivu*. In 1949, he joined the *Service géologique du Congo belge*, and became involved in problems of applied geology related to the Kivu Province. By the end of 1951 and in early 1952, Meyer witnessed a volcanic eruption on the northern flanks of the Nyamuragira volcano (Kivu). His professsional duties subsequently brought him to the Eastern Province of the former Belgian Congo, where he was involved in projects on hydrology and civil engineering.

The volcanological observations on the 1951-1952 eruption of the Nyamuragira published by Meyer in 1953 caught the attention of Victor Van Straelen, who then was the head of the Institute of the National Parks of the Belgian Congo (IPNCB) in Brussels. In 1955, Meyer was engaged by the IPNCB. His main task was to study volcanism in Kivu, and in particular in the Virunga region at the borders of the modern Democratic Republic of the Congo, Uganda and Rwanda. Unfortunately, the Congolese declaration of independence on June 30, 1960, forced him to leave the Kivu area. He moved to Léopoldville (present-day Kinshasa), where he served as an advisor to the *Ministère de la Géologie et des Mines du Congo* for two years. After returning to Belgium, Meyer left for Africa again, where he worked as an UN-expert in Gabon. As a result of the complications following an attack of viral haepatitis, he died in Libreville, the capital of this country, on January 10, 1965, at age of 44.

André Meyer's publication record is essentially devoted to igneous rocks from Kivu, in particular to lavaflows of the Nyiragongo and Nyamuragira volcanoes, as well as to the rocks forming the syenite-carbonatite complex of the Lueshe (papers often written in co-authorship with Pierre de Béthune).

Biography

Mortelmans, G. (1965) *Bulletin de la Société belge de Géologie, de Paléontologie et d'Hydrologie*, **74**, 9-10.

Bibliography

Mortelmans, G. (1965) *Bulletin de la Société belge de Géologie, de Paléontologie et d'Hydrologie*, **74**, 10-11.

ANTHOINITE

Raymond Anthoine (1888-1971).

$AlWO_3(OH)_3$

Triclinic.

Mount Misobo, Kalima, Maniema Province, Democratic Republic of the Congo.

Belgium: Royal Museum for Central Africa, Tervuren - Type: CT N° 2426.

IMA: G

Varlamoff, N. (1947) Anthoinite, nouveau tungstate hydraté d'alumine. *Annales de la Société géologique de Belgique*, **70**, B153-166.

Sahama, Th.G. (1981) The secondary tungsten minerals, a review. *The Mineralogical Record*, **12**, 81-87.

http://www.handbookofmineralogy.org/pdfs/anthoinite.pdf

Anthoinite was first found in 1943 in tin- and tungsten-rich eluvial deposits in the Maniema Province of the former Belgian Congo. In 1946 fresh material was collected when trenches were cut into quartzite veins containing tungsten ore. The large crystals of anthoinite, described and shown by Varlamoff (1947) are pseudomorphs after scheelite (Sahama, 1981).

Cream-coloured anthoinite.
Bugambira, Minetain, Rwanda.
Royal Belgian Institute of Natural Sciences (N° B7750).

Anthoinite in quartz.
Nyakabingu Mountain, Bumbogo, Rwanda.
Royal Belgian Institute of Natural Sciences (N° B7737).

Raymond Anthoine was born on May 12, 1888, in Marcinelle (Charleroi), in the Hainaut Province. He enrolled at the University of Liège (ULg), where he graduated as a mining engineer in 1911. He continued his studies at the same institution, and earned an electromechanical engineering degree one year later followed by a diploma of engineering geologist in 1919. In 1912, Anthoine accepted to work for the *Société française des Mines de Fer*, in France. In this function he had to conduct and control subsurface exploration activities and geological mapping. He also had to estimate the reserves of the ore deposits of the different concessions of the company in the Manche department.

At the close of the First World War and until 1920 Anthoine served as an assistant to the geology department of the ULg. At about the same time he became a consultant to the *Régie Industrielle des Mines d'Or de Kilo-Moto*. This Belgian state entity was created in 1919 by the Belgian government to administer the gold mining at Kilo and in the Moto River basin, in the northeast of the former Belgian Congo; it was extended in 1926 to form the *Société des Mines d'Or de Kilo-Moto*. Over the years, Antoine subsequently became director, managing director, administrator and member of the Board of this company. Meanwhile, he accomplished many scientific missions in southern Europe, the Balkans and Africa, in particular in the former Belgian Congo, South Africa, Rhodesia (now Zambia and Zimbabwe), Nyassaland (present-day Malawi), and Mozambique. These missions generally focused on mining activities, geological mapping and ore extraction operations.

Anthoine was a founding member of the *Syndicat Minier Africain* (SYMAF) (1929) and the *Syndicat Minier d'Etain* (SYMETAIN) (1932). He also played a prominent part in the foundation of the *Société minière et géologique du Zambèze* (1920). As a consultant to this company, which was involved in projects in Mozambique and other regions abuting on the Zambezi River, he was involved in the exploration, mining and exploitation of ore deposits. He always paid attention to gold prospection as well. Anthoine died in Brussels on June 4, 1971.

Biography

Staner, P. (1971) *Mededelingen der Zittingen van de Koninklijke Academie voor Overzeese Wetenschappen/ Bulletin des Séances de l'Académie royale des Sciences d'Outre-Mer*, **XVII**, 666-667.

Records available at the Royal Academy of Overseas Sciences (KAOW/ARSOM) in Brussels.

Lucien Bia (1852-1892).

Mushishimano, Haut-Katanga Province, Democratic Republic of the Congo.

IMA: D

Buttgenbach, H. (1927-28) Notes sur la bialite, nouveau minéral, et sur quelques autres minéraux du Katanga. *Annales de la Société géologique de Belgique*, **51**, C117-120.
Embrey, P.G. & Fejer, E.E. (1969) Tavistockite and bialite discredited. *Mineralogical Magazine,* **37**, 123-127.

Emile Richet provided the specimen Buttgenbach analysed: spherules composed of white needles on a dense, phosphate matrix. In 1969 Embrey & Fejer demonstrated that *bialite* was identical with wavellite.

Lucien Bia was born in Liège on December 2, 1852. When France and Germany declared war on each other in July 1870, he joined the Belgian Army and became lieutenant of the 2nd Guides Regiment in the early 1880s. Being drawn to an adventurous lifestyle, he embarked for the Belgian King Leopold II's Congo Free State (CFS) on March 15, 1887, where he would stay for about three years. Because of his great military merits and African experience, he was appointed captain upon his return to Belgium (1890). Shortly after, on Henry Morton Stanley's recommendation, he was invited by the *Comité du Katanga* to lead an expedition to southern Katanga. In order to fulfill this task succesfully, Bia was assisted by lieutenant Emile Francqui (second-in-command), Jules Cornet (third-in-command) (*see*: cornetite), Jules Amerlinck (expedition doctor), and lieutenant Eugène Derscheid. They all boarded the vessel *S/S Africa* in Antwerp on May 18, 1891, destination Boma, on the Atlantic coast of the CFS.

Early 1892, the Bia Expedition reached Bunkeya (southeastern Katanga), the capital of the Yeke Kingdom, that was founded and ruled at the time by King Mwenda Msiri. It was not far away from this city that, in late December 1891, Msiri was shot and killed by the Belgian officer Omer Bodson, a member of the Stairs Expedition to Katanga (1891-1892). Whilst Cornet concentrated his geological observations on the copper deposits of southern Katanga, Bia and Francqui explored the surroundings of the Luapula River. They agreed that all members of the expedition were to meet at Tenke, a locality on the banks of the Lufira River. During his reconnaissance work, Bia fell victim of a liver disease. This illness, in combination with exhaustion and hardships, finally took its toll on August 30, 1892, when he died at Tenke. Bia was buried near Tenke, on top of the Ditakata Hill, and Francqui succeeded him as leader of the expedition.

Biography

Buttgenbach, H. (1951) *Biographie coloniale belge*, **II,** col. 58-62.

BILLIETITE

Valère Billiet (1903-1945).

$Ba(UO_2)_6O_4(OH)_6.8H_2O$

Orthorhombic.

Shinkolobwe, Haut-Katanga Province, Democratic Republic of the Congo.

USA: Harvard University, Cambridge, Massachusetts, N° 104455.
USA: National Museum of Natural History, Washington, DC, N° 160496.

IMA: G

Vaes, J.F. (1947) Six nouveaux minéraux d'urane provenant de Shinkolobwe (Katanga) : la billietite (urane hydraté de Ba), la vandendriesscheite (uranate hydraté de Pb), la masuyite (uranate hydraté de Pb), la richetite (uranate hydraté (?) de Pb), la studtite (carbonate hydraté d'urane) et la diderichite (carbonate hydraté d'urane). *Annales de la Société géologique de Belgique,* **70**, B212-225.

Schoep, A. & Stradiot, S. (1948) Additional data on the properties of becquerelite and billietite. *The American Mineralogist*, **33**, 503-507.

Thoreau, J. (1948) Caractères cristallographiques de la billietite et la vandendriesscheite. *Annales de la Société géologique de Belgique,* **71**, B76-78.

Brasseur, H. (1949) Etude de la billietite. *Bulletin de l'Académie royale de Belgique, Classe des Sciences, série 5,* **35**, 793-804.

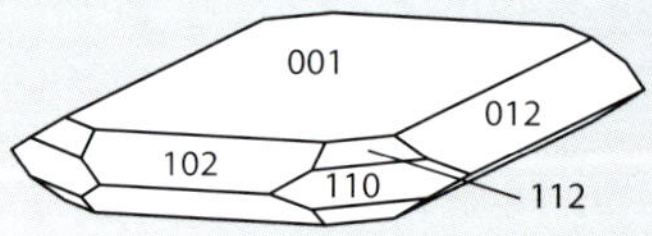

after Vaes (1947)

http://www.handbookofmineralogy.org/pdfs/billietite.pdf

After the discovery and first description of billietite by Vaes (1947), Schoep & Stradiot (1948) considered billietite as a Ba-bearing variety of becquerelite. At about the same time, Thoreau (1948) and Brasseur (1949) pointed out that billietite was indeed a new mineral species and not a variety of another mineral.

Crust of small billietite crystals.
Shinkolobwe, Haut-Katanga Province, D.R. Congo.
J.-C. Leydet.

Tabular crystal of billietite.
Shinkolobwe, Haut-Katanga Province, D.R. Congo.
L. Vandenberghe.

Tabular crystals of billietite.
Shinkolobwe, Haut-Katanga Province, D.R. Congo.
Royal Belgian Institute of Natural Sciences (N° C 4912).

Short-prismatic brown orange billietite crystals on yellow studtite.
Shinkolobwe, Haut-Katanga Province, D.R. Congo.
F. Coune (N° 3025).

Crust of billietite and grey rutherfordite nodules.
Shinkolobwe, Haut-Katanga Province, D.R. Congo.
J.-C. Leydet.

Valère Billiet was born in Ghent, in the East-Flanders Province, on February 14, 1903. He entered the local university, where he received his doctor's degree in mathematics and physics in 1924. After graduating, he volunteered at the laboratory of mineralogy and crystallography of the Faculty of Mathematics and Natural Sciences of Ghent University (GU), headed by Alfred Schoep (*see*: schoepite). Here, he did experimental work, improved his theoretical knowledge of X-ray analysis, and in his spare time studied problems of geometrical crystallography and descriptive mineralogy. To obtain financial independence, Billiet was forced to lecture at secondary schools in Aalst, Tongeren and Ghent.

At the start of his professional career, Billiet's main scientific concern was the study of refractive indices of crystals with high refractive power. Several of the mineral species he examined during this period were discovered shortly before and first described by Schoep. Subsequently, Billiet paid attention to other physical properties of a variety of minerals as well. At the same time he did his utmost best in order to acquire a röntgen-analysis instrument to serve the research needs of the scientific community in Ghent. Despite financial hardships, he spent the summer holidays of 1931 and 1932 abroad to familiarize himself with the röntgenographic technique. At the *Technische Hogeschool* (Technical University) of Delft, in The Netherlands, he worked with Wieger Fokke de Jong and Jan Adolf Grutterink. At the Faculty of Sciences in Paris he worked under the French mineralogist Charles Mauguin. In the fall of 1932, Billiet was admitted to the laboratory of mineralogy and crystallography of GU where he became an assistant. In this position, he finally succeeded to set up a laboratory of röntgen-analysis at the Ghent alma mater. From then on, his field of research extended considerably and he worked closely together with Adriaan Vandendriessche (*see*: vandendriesscheite). Both scientists were the very first to apply radiocrystallographic techniques to minerals from the former Belgian Congo. In 1941, because of his expertise on irradiation phenomena, radioactivity and X-ray analysis, Billiet became a member of the academic staff of the Faculty of Medicine of GU, where he was in charge of teaching and cancer research.

During World War II, Billiet was active in the resistance movement and served as a member of the undercover intelligence service. In August 1944, he was arrested and eventually transferred to the concentration camp of Neuengamme (Germany). Towards the end of April 1945 over 5,000 prisoners brought down from Neuengamme were forced to board the *SS Cap Arcona,* an ocean liner of the Hamburg South America line at anchor in the Bay of Neustedt (Lübeck). On May 3, 1945 - hardly a few days before the end of the war - the *Cap Arcona*, including Billiet and all its other prisoners, was bombed by airplanes of the Royal Air Force. Shortly before the ship sank, Billiet was shot by an SS officer because he was distributing life-buoys to sick and needy people on board.

Biography

Hacquaert, A. (1946) In memoriam Valère Billiet en Karel Verlat. In: Verslag over den toestand van de Universiteit gedurende het academisch jaar 1944-1945, 36-51. Gent: Universiteit Gent.
Rijksuniversiteit te Gent. Liber memorialis 1913-1960, **II**, 324-326, 1960. Gent: Uitgave van het Rectoraat.
Walschot, L. (1998) *Natuurwetenschappelijk Tijdschrift*, **78**, 294-295.

Bibliography

Rijksuniversiteit te Gent. Liber memorialis 1913-1960, **II**, 326-327, 1960. Gent: Uitgave van het Rectoraat.

BLATONITE

Norbert Blaton, (°1945).

$UO_2CO_3.H_2O$

Hexagonal / Trigonal.

Jomac Mine, Brown's Rim, White Canyon district, San Juan County, Utah, USA.

Belgium: Royal Belgian Institute of Natural Sciences, Brussels - Type: HT: RC 4788.

IMA N°: 1997-025

Vochten, R. & Deliens, M. (1998) Blatonite, $UO_2CO_3.H_2O$, a new uranyl carbonate monohydrate from San Juan County, Utah. The Canadian Mineralogist, **36**, 1077-1081.

http://www.handbookofmineralogy.org/pdfs/blatonite.pdf

SEM photograph of blatonite (overview). Jomac Mine, San Juan County, Utah, USA. Royal Belgian Institute of Natural Sciences.

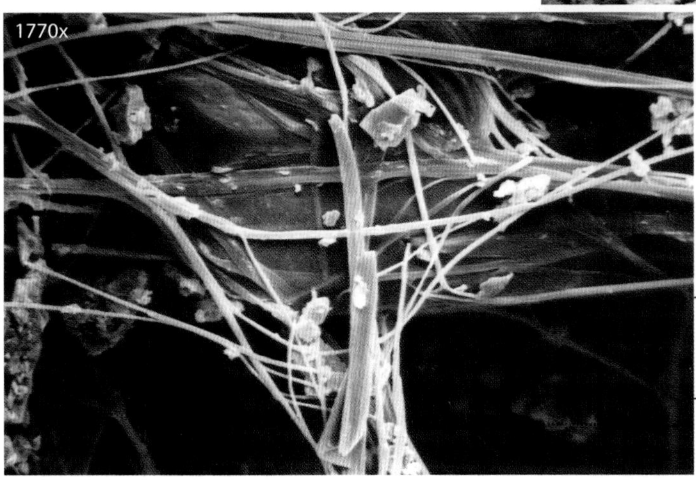

SEM photograph of fibrous blatonite. Jomac Mine, San Juan County, Utah, USA. Royal Belgian Institute of Natural Sciences.

Pointed fusiform yellow blatonite with gypsum.
Jomac Mine, San Juan County, Utah, USA.
R. Vochten (N° 524514).

Acicular yellow blatonite.
Jomac Mine, San Juan County, Utah, USA.
Royal Belgian Institute of Natural Sciences (N° C5080).

Fusiform yellow blatonite ending in a needle.
Jomac Mine, San Juan County, Utah, USA.
Royal Belgian Institute of Natural Sciences (N° C4974).

Thick needle-shaped blatonite crystal with rough surface.
Jomac Mine, San Juan County, Utah, USA.
Royal Belgian Institute of Natural Sciences (N° C4974).

Yellow blatonite crystals with white gypsum crust.
Jomac Mine, San Juan County, Utah, USA.
Royal Belgian Institute of Natural Sciences (N° C4974).

Norbert Blaton was born on May 25, 1945 in Sint-Maria-Latem (Zwalm), in the East-Flanders Province. Following his secondary education in Oudenaarde and an additional year in Ghent, he enrolled at the Catholic University of Leuven (CUL), where he earned the degree of licentiate in chemistry in 1968. From 1968 until 1970, he held a grant of the *Institute for the Encouragement of the Scientific Research in Industry and Agriculture* (IWONL-IRSIA) in Brussels. As such, he worked at the Faculty of Pharmaceutical Sciences of the *Katholieke Universiteit Leuven* (KUL). At this faculty, he submitted a PhD thesis commissioned by *Métallurgie Hoboken* dealing with the separation of cobalt and nickel applied to metallurgy. After obtaining his PhD in 1974, Blaton spent the rest of his professional career at the KUL, going through all the scientific and academic ranks. He started out as an assistant (1970-1975) and finally ended his career by taking on the responsibilities of a full professorship. He held this position from 2001 onward, up till he became professor emeritus. Upon his retirementon October 1, 2007, he was head of the laboratory of biocrystallography of the KUL.

Throughout his scientific and academic career, the research of Blaton mainly concentrated on: (a) structural analysis of ligands and metallic complexes related to analytical methods; (b) qualitative, quantitative and structural analysis of crystalline powder samples using powder diffraction, in support of the production of medicines; (c) structural analysis of hydrogen-bridge forming bonds related to spectroscopic studies, and, last but not least, (d) crystallographic studies in relation to mineralogy, including the analysis of iron-bearing minerals and determination of the structure and other properties of minerals using powder diffraction and monocrystal techniques. Within the framework of the latter research topic, Blaton played an important part in the discovery and circumstantial description of two new secondary uranium minerals (*see*: piretite and deliensite).

BOODTITE -> impure HETEROGENITE

Anselmus Boetius De Boodt (1550-1632).

Star of the Congo Mine (l'Etoile du Congo Mine), Lubumbashi, Haut-Katanga Province, Democratic Republic of the Congo.

IMA: D

De Leenheer, L. (1936) Nieuwe Kobalt-mineralen. *Natuurwetenschappelijk Tijdschrift*, **18**, 77-78.

Billiet, V. & Vandendriessche, A. (1940) Les oxydes hydratés de cobalt du Katanga. *Bulletin de la Société belge de Géologie, de Paléontologie et d'Hydrologie*, **49**, 63-78.

Hey, M.H. (1962) Cobaltic hydroxide in nature. *Mineralogical Magazine*, **33**, 253-259.

De Leenheer (1936) presented the results of his analyses of three cobalt-bearing specimens. He concluded that one of them was a new mineral species and named it *boodtite.*
Investigations by Billiet & Vandendriessche (1940) proved that *boodtite*, and another new Co-mineral described by De Leenheer (*see*: *trieuite*) were impure varieties of heterogenite. This conclusion was confirmed by Hey (1962).

Anselmus Boetius De Boodt was born in the mid-16th century (c. 1550) in Bruges. He decided to study law at the universities of Leuven and Douai, attended lectures of church and civil affairs in Orléans, and studied medicine in Heidelberg under Thomas Erastus. This Swiss theologian was an expert in rocks from both a scientific and a medical point of view. During the rule of Rudolf II, grandson of Charles V, De Boodt stayed at the Austrian Court (Vienna) in the early 1580s. When Rudolf II (who had a great interest in arts and natural sciences) moved his court to Prague, De Boodt accompanied him and took this opportunity to improve his knowledge of alchemy and medicine.

In 1586 De Boodt temporarily left the Austrian imperial household and attended the University of Padua. After obtaining his doctor's degree in medicine (1587), he returned to Prague. Here, the young physician and imperial lapidarius prepared his *Historia Naturalis,* and a *Gemmarum et Lapidum Historia*. This fundamental manual on mineralogy and gemology is still considered as the most important lapidary of the 17th century today. De Boodt's lapidary was published almost simultaneously at Hanau and Frankfurt (1609). It presented different aspects of precious stones and fossils, including their medical applications, and comprised the first attempt at a systematic description and classification of rocks since the work of the Roman naturalist and natural philosopher Pliny the Elder. Several reprints of this book are known. It was also translated in French by Jean Bachou from Lyon and published under the title of *Le parfait joallier.* De Boodt's work was valued to such a degree that he was appointed adviser and court physician to the emperor.

After Rudolf II's death, De Boodt returned to Bruges where, from 1615 until his death, he settled in the parental house at the Ridderstraat. Here he compiled a herbarium. This work had not yet been published when he passed away in his native town on June 21, 1632.

Biography

Dewalque, G. (1873) *Biographie nationale*, **4**, col. 814-816.

De Boodt-Maselis, M.-C. (1981) Anselmus Boetius De Boodt (Brugge 1550-1632). Een Vlaams humanist met Europese faam. Gevolgd door een reproductie van het kruidenboek (1640) met Nederlandse vertaling door E.H. De Schrijver, 99 p. Handzame: Familia et Patria.

De Boodt-Maselis, M.-C. (1981) *Nationaal biografisch woordenboek,* **9**, col. 72-79.

De Boodt-Maselis, M.-C. (2002) Anselmus Boetius De Boodt (1550-1632) Brugs humanist aan het Hof van Oostenrijk. *Kunsttijdschrift Vlaanderen*, **51/1**, 21-22.

Bibliography

De Boodt-Maselis, M.-C. (1981) Anselmus Boetius De Boodt (Brugge 1550-1632). Een Vlaams humanist met Europese faam. Gevolgd door een reproductie van het kruidenboek (1640) met Nederlandse vertaling door E.H. De Schrijver, 87-89. Handzame: Familia et Patria.

BORGNIEZITE -> sodium AMPHIBOLE

Georges Borgniez (1901-1965).

Lueshe, Nord-Kivu Province, Democratic Republic of the Congo.

IMA: D

de Béthune, P. & Meyer, A. (1956) Les carbonatites de la Lueshe (Kivu, Congo belge). *Comptes rendus de l'Académie des Sciences, Paris*, **243**, 1132-1134.
Fleischer, M. (1957) New Mineral Names. *The American Mineralogist*, **42**, 440-444.
Leake, B.E. (1978) Nomenclature of amphiboles. *The American Mineralogist*, **63**, 1023-1052.

One year after the first description of borgniezite by de Béthune & Meyer (1956), Fleischer (1957) wrote: "there is no excuse for burdening the literature with such names".
The name was officially discredited by Leake (1978) in the first report on amphibole nomenclature.

Georges Borgniez was born on March 6, 1901, in Tienen, in the Flemish Brabant Province. Following his secondary education at Thuin (1913-1919), he enrolled at the *Ecole Polytechnique de Mons.* He graduated as a mining engineer from this institution in 1924. Following the completion of his military service (1924-1925), Borgniez started his professional career by doing geological work in Belgium and France. From 1926 till 1928 he joined the *Comité Spécial du Katanga* in the former Belgian Congo. At the end of this two year stay in central Africa, He shortly returned to Belgium and started working for the *Société Belge de Recherches Minières en Afrique* (REMINA). In this capacity he returned to the former Belgian Congo during the fall of 1928. It was the start of a twenty year long collaboration with this colonial society, and a lifelong and deeply cherished association, co-operation and friendship with Fernand Delhaye (*see*: delhayelite).

The economic crisis of the early 1930s allowed Borgniez to pursue studies at the Sorbonne and the *Musée national d'Histoire naturelle* in Paris. It also allowed him to spend about one year (July 1931-September 1932) in the former French colony of Equatorial Africa and Cameroon. During the academic year 1932-1933, Borgniez attended courses of geology at the *Ecole Polytechnique de Mons*, where he earned his engineering geology degree in 1933. Subsequently, he returned to French Equatorial Africa and Cameroon, and remained there until 1935. With the exception of a rather long period of convalescence in Belgium and two holiday travels to South Africa, Borgniez served in Africa from November 1935 up till the close of the Second World War. From 1948 to the late 1950s his scientific missions focused on hydrological projects in the former Belgian Congo and Ruanda-Urundi. Here he worked for the *Régie de Distribution d'Eau* (REGIDESO), the hydrological service in charge of the local water-supply. His career in the former Belgian Congo came to an end following the countries independence in late June, 1960. Borgniez passed away at Auderghem (Brussels) on February 2, 1965.

Biography

Anonymous (1965) Jules Cornet 1865-1929. Centenaire de sa naissance 1965, 185-187. Bruxelles: Robert Louis.

Lederer, A. (1973) *Biographie belge d'Outre-Mer*, **VII-A**, col. 58-62.

Bibliography

Anonymous (1965) Jules Cornet 1865-1929. Centenaire de sa naissance 1965, 187-188. Bruxelles: Robert Louis.

Lederer, A. (1973) *Biographie belge d'Outre-Mer*, **VII-A**, col. 62-63.

BRIARTITE

Gaston Briart (1897-1962).

Cu_2FeGeS_4

Tetragonal.

Prince Leopold Mine, Kipushi, Haut-Katanga Province, Democratic Republic of the Congo.

France: Musée de Minéralogie, Ecole des Mines de Paris - Type: HT.
United Kingdom: National History Museum, London - Type: CT: 1967271.
Belgium: *Université Catholique de Louvain*, Louvain-la-Neuve - Type: HT: U 309.

IMA N°: 65-018

Francotte, J., Moreau, J., Ottenburgs, R. & Lévy C. (1965) La briartite, $Cu_2(Fe,Zn)GeS_4$, une nouvelle espèce minérale. *Bulletin de la Société française de Minéralogie et de Cristallographie*, **88**, 432-437.

http://www.handbookofmineralogy.org/pdfs/briartite.pdf

Massive briartite with metallic lustre.
Kipushi, Haut-Katanga Province, D.R. Congo.
P. De Bondt.

Massive briartite with renierite.
Kipushi, Haut-Katanga Province, D.R. Congo.
R. Vandenbosch.

BRIARTITE

Gaston Briart was a grand-nephew of the renowned engineering geologist Alphonse Briart, a very close friend and colleague of mining engineer François-Léopold Cornet, the father of Jules (*see*: cornetite). Gaston Briart was born on October 26, 1897. Following the completion of his classical studies, he attended courses at the *Ecole Polytechnique de Mons* and obtained his mining engineer degree in 1924. On February 1, 1926, he started working in Africa for the *Union Minière du Haut-Katanga* (UMHK) and for many years he served the geological department of this company. A substantial part of Briart's professional activities focused on the study of the rock formations exposed in the Prince Leopold Mine (now Kipushi Mine), Kipushi, Katanga, in the former Belgian Congo. After the Second World War (1946), he left Africa for Belgium. Here he continued his activities for UMHK in Brussels, initially as an engineering geologist and finally in the position of chief engineer of the department of geology. UMHM consultant Gaston Briart passed away in Brussels on June 13, 1962, at the age of 64.

Biography

Anonymous (1965) Jules Cornet 1865-1929. Centenaire de sa naissance 1965, 237. Bruxelles: Robert Louis.

Bibliography

Anonymous (1965) Jules Cornet 1865-1929. Centenaire de sa naissance 1965, 237. Bruxelles: Robert Louis.

BUTTGENBACHITE

Henri Buttgenbach (1874-1964).

$Cu_{36}(NO_3)_2Cl_6(OH)_{64}.nH_2O$

Hexagonal.

Likasi Mine, Likasi, Haut-Katanga Province, Democratic Republic of the Congo.

Belgium: Laboratory of Mineralogy, University of Liège - Type: CT: 9482 and 16915

IMA: G

Schoep, A. (1925) Sur la buttgenbachite, nouveau minéral. *Comptes rendus de l'Académie des Sciences, Paris*, **181**, 421-422.
Buttgenbach, H. (1926-27) Cristaux de connellite-buttgenbachite. *Annales de la Société géologique de Belgique,* **50**, B35-40.

http://www.handbookofmineralogy.org/pdfs/buttgenbachite.pdf

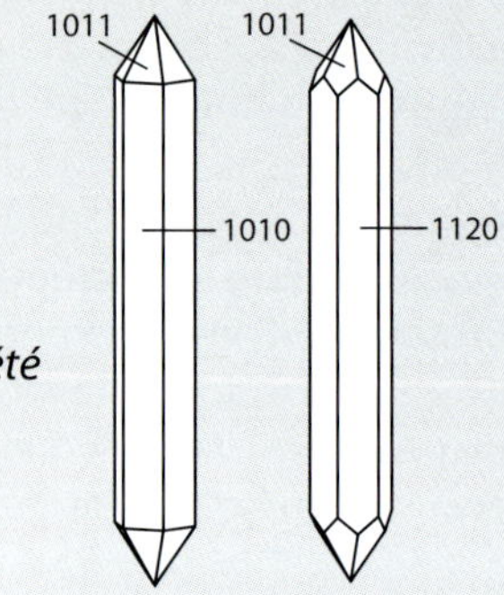

after Buttgenbach (1926-27)

Aggregate of blue acicular buttgenbachite on cuprite with pale malachite coating.
Likasi, Haut-Katanga Province, D.R. Congo.
Royal Museum for Central Africa (N° 11431).

Aggregate of blue acicular buttgenbachite.
Likasi, Haut-Katanga Province, D.R. Congo.
G. Gauthier.

Long-prismatic buttgenbachite.
Likasi, Haut-Katanga Province, D.R. Congo.
Royal Museum for Central Africa (N° 4217).

Long-prismatic blue buttgenbachite
on red cuprite.
Likasi, Haut-Katanga Province, D.R. Congo.
J. & N. Mornie.

Henri Buttgenbach was born in Ensival (Verviers), in the Liège Province, on February 5, 1874. In 1892, he enrolled as a pre-engineering student at the University of Liège (ULg). At this institution he obtained his candidate degree of mining engineering in 1894 and that of sciences in 1896. As a result of his great interest in mineralogy, he attended lectures given by Giuseppe Cesàro (*see*: cesàrolite) during his candidate studies. When Cesàro became aware of the talents and intelligence of his young pupil, he engaged him as a student-assistant (1895), a position he held until 1899. Then, instead of completing his mining engineering studies, Buttgenbach left his homeland and travelled to several countries worldwide.

In 1902, Buttgenbach was engaged by the *Comité Spécial du Katanga*. He was to join an expedition headed by George Grey and organized by the British Tanganyika Concessions Ltd. Company (TCL). This expedition was in charge of prospecting for valuable minerals in Upper Katanga (former Belgian Congo). One of the goals of Buttgenbach's participation to this field trip was to control by stealth the exploration activities of his British colleagues. On June 16, 1902, Buttgenbach met and joined the TCL exploration team and during the next couple of years he visited the main gold and copper mines of Katanga. Subsequently, he returned to Belgium. Two years later (1906) he returned to the former Belgian colony and was employed in the gold district of Kilo-Moto, Ituri, in the northeastern part of this country.

At the establishment of the *Union Minière du Haut-Katanga* (UMHK) (October 1906), Buttgenbach was appointed member to the Board. From 1907 to 1912 he acted as managing director of this company. In June 1911, he witnessed the smelting of copper, a procedure carried out for the first time in the history of UMHK. This took place in Elisabethville (modern Lubumbashi). Buttgenbach took benefit of World War I to publish a treatise entitled *Les minéraux et les roches* (1917). During the Interbellum, he stayed in Belgium and did not set foot in central Africa. When Cesàro retired from the ULg in 1921, Buttgenbach was appointed full professor in charge of teaching crystallography and mineralogy. In 1925, he published a memoir entitled *Minéralogie du Congo belge* followed in 1947 by *Les minéraux de Belgique et du Congo belge*. Both are nowadays still valuable as reference works on this subject.

During his professional career, Buttgenbach discovered and described several new minerals. Some are still standing as valid species and are illustrated in the present work (cesàrolite, fourmarierite, thoreaulite). Buttgenbach retired as an emeritus professor in 1945. However, he remained interested in the former Belgian Congo and was able to return there twice more. When the UMHK celebrated its fifty years of existence (1956), he was the last surviving member of the companies first board. Buttgenbach passed away at his home in Sint-Pieters-Woluwe (Brussels) on April 29, 1964, aged 90. He is said to be the father of Congolese mineralogy.

Biography

Thoreau, J. (1966) *Annuaire de l'Académie royale de Belgique*, **CXXXII,** 125-138.
Mélon, J. (1968) *The American Mineralogist* **53**, 536-539.
Aderca, B. (1973) *Biographie belge d'Outre-Mer*, **VII**-A, col. 77-80.
Lepersonne, J. (1981-1982) *Biographie nationale*, **42**, col. 109-116.

Bibliography

Thoreau, J. (1966) *Annuaire de l'Académie royale de Belgique*, **CXXXII,** 138-160.
Mélon, J. (1968) *The American Mineralogist* **53**, 539-544 (selected papers).
Aderca, B. (1973) *Biographie belge d'Outre-Mer*, **VII**-A, col. 81-82 (only papers in relation to the former Belgian Congo).

CATTIERITE

Félicien Cattier (1869-1946).

CoS_2

Cubic.

Shinkolobwe Mine, Haut-Katanga Province, Democratic Republic of the Congo.

IMA: G

Kerr, P.F. (1945) Cattierite and vaesite: new Co-Ni minerals from the Belgian Congo. *The American Mineralogist*, **30**, 483-497.

http://www.handbookofmineralogy.org/pdfs/cattierite.pdf

Bronze-coloured cattierite crystals.
Shinkolobwe, Haut-Katanga Province, D.R. Congo.
P. De Bondt.

Grey cattierite crystals with metallic luster.
Shinkolobwe, Haut-Katanga Province, D.R. Congo.
R. van Dooren.

Félicien Cattier was born on March 4, 1869, in Cuesmes (Mons), in the Hainaut Province. Here, his father was schoolmaster and had among his pupils Jules Cornet (*see*: cornetite) and Charles Lemaire (*see*: dewindtite). Both Cornet and Lemaire befriended the young Félicien during their childhood. Years later, like Cattier, they were involved in the exploration and economic development of the Congo Free State (CFS). After his secondary schooling, Cattier enrolled at the Free University of Brussels (BFU), graduating as Doctor in Law, and Doctor in Political and Administrative Sciences during the early 1890s. He practiced under Edmond Picard, a top lawyer at the Court of Appeal and the Court of Cassation of Belgium, and highly devoted to colonial affairs. Early on in his career, Cattier served at the BFU, being promoted to extraordinary professor in 1900 and to full professor in 1906. In the meantime, he established relations with the business-community. In 1918, he abandoned academic life but continued, whenever possible, to render to good services the Brussels alma mater.

The first contacts of Cattier with the economic sector date back to 1900. That year, thanks to the Belgian colonial pioneer and businessman Albert Thijs (who was very active in the CFS and well acquainted with King Leopold II of Belgium), he was appointed secretary-general to the *Compagnie Internationale d'Orient*, a company much engaged with investments in China. Among his new colleagues, Alexandre Delcommune (*see*: *diderichite*) and Emile Francqui (*see*: cornetite) are to be mentioned. Bearing such great responsibilities Cattier could rely on the experience he had gained serving as a juridical adviser during a mission to Siam (modern Thailand) years before. Still prior to the outbreak of the World War I, Cattier accepted to become managing director of the *Banque d'Outremer* and as such had some pull in several societies. After the take-over of this financial organization by the *Société Générale de Belgique* (SGB) (1928), Cattier held a high-rank managerial position in the business-community. He headed several important colonial societies linked to the SGB. Through his daily contacts with these societies, Cattier's knowledge of and interest in the former colony of Belgium increased continuously, and important missions were left in his charge by the Belgian government. From time to time, he also acted as a representative of Belgium at international meetings and conferences.

Upon the death of Emile Francqui in 1935, Cattier became president of the University Foundation in Brussels, the Belgian National Fundfor Scientific Research (NFWO-FNRS), the National Cancer Fund and the Francqui Foundation. Cattier untimely passed away on February 6, 1946, on Madeira Island (Portugal).

Biography

Louwers, O. (1947) *Bulletin des Séances de l'Institut Royal Colonial Belge*, **18**, 111-126.

Vanzype, G. (1950) *Annuaire de l'Académie royale de Belgique*, **CXVI**, 125-135.

Kauch, P. (1964) *Biographie nationale*, **32**, col. 90-95.

Louwers, O. (1967) *Biographie belge d'Outre-Mer*, **VI**, col. 189-201.

Kurgan-van Hentenrijk, G., Jaumain, S. & Montens, V. (eds) (1996) Dictionnaire des Patrons en Belgique. Les Hommes, les Entreprises, les Réseaux, 96-98. Bruxelles: De Boeck & Larcier, Dept. De Boeck Université.

Bibliography

Vanzype, G. (1950) *Annuaire de l'Académie royale de Belgique* **CXVI**, 136 (selected papers).

Louwers, O. (1967) *Biographie belge d'Outre-Mer,* **VI**, col. 201.

CESÀROLITE

Giuseppe Cesàro (1849-1939).

$Pb(Mn^{4+})_3O_6(OH)_2$

Hexagonal.

Sidi Amor ben Salaam, Tunisia.

France: Muséum national d'Histoire naturelle, Paris - Type: T: 120275 (donated by Henri Buttgenbach).
Belgium: Laboratory of Mineralogy, University of Liège - Type: HT.

IMA: G

Buttgenbach, H. & Gillet, C. (1920) La cesàrolite (nouvelle espèce minérale). *Annales de la Société géologique de Belgique*, **43**, B239-241.

http://www.handbookofmineralogy.org/pdfs/cesarolite.pdf

SEM photograph of botryoidal cesàrolite.
Sidi Amor ben Salaam, Tunisia.
Laboratory of Mineralogy, University of Liège.

Globular crust of dark grey cesàrolite.
Sidi Amor ben Salaam, Tunisia.
Royal Belgian Institute of Natural Sciences (B 4157).

Globular crust of dark grey cesàrolite, partially corroded.
Sidi Amor ben Salaam, Tunisia.
Royal Belgian Institute of Natural Sciences (B 4157).

Coralloidal cesàrolite.
Sidi Amor ben Salaam, Tunisia.
H. Dillen (N° 4.4.6.10/1).

Cross-section through a globular cesàrolite aggregate with hemimorphite.
Mt. Chemin, Valais, Switzerland.
F. Hubert (N° 11447).

Giuseppe Cesàro was a native of Naples, in the former Kingdom of the Two Sicilies. He was born on September 7, 1849. He completed his early schooling in his native town. In 1865, upon the advice of a friend of his father, he was sent to Belgium, where his first priority was to make himself conversant with the French language. Subsequently, he enrolled as a mining engineer student at the *Ecole des Mines* in Liège. However, as a result of family misfortunes, bad health and other adverse events he had to stop his studies after finishing his third year. To earn his livelihood, he decided to give private lessons in mathematics, chemistry and physics.

These unfavourable living conditions lasted nearly fifteen years. In his spare time Cesàro nevertheless managed to attend lectures on crystallography and mineralogy at the University of Liège (ULg) given by Gustave Dewalque (*see*: *dewalquite*). Indeed, ever since his childhood he had developped a deep interest and love for crystals and minerals. Most of them he had discovered in the vicinity of Naples and originated from the nearby Vesuvius volcano. The long interruption of his study of engineering, however, caused Cesàro to abandon his intention of resuming his studies. Instead, after consulting with Dewalque, he preferred to start a thorough study of the mineral collection of the Liège alma mater. Through self-study acquired such an in-depth knowledge of mineralogy that as early as 1883 the *Annales de la Société géologique de Belgique* accepted to publish some of his writings. Cesàro's publications were of such high standard that, when Dewalque relinquished the classes of mineralogy and crystallography (1891), he was put in charge of both courses at the university, despite the fact that he did not have the required certificates of qualification. His Italian origin did not hamper his appointment either, as he obtained Belgian citizenship as early as 1888. It was the start of a most brilliant career at the ULg that lasted about thirty years. In 1895 he was appointed extraordinary professor and full professor in 1900. Among the mineral species first described by Cesàro and discussed in this book are cornetite, fraipontite, koninckite and richellite (in collaboration with G. Despret).

In 1919, Cesàro obtained the emeritus status. He was succeeded by Henri Buttgenbach (*see*: buttgenbachite), one of his students. After retiring from active service and nearly up to the 1930s, Cesàro continued to work at the university. He passed away at his country house in Comblain-au-Pont, near Liège, on January 20, 1939. Cesàro was a good friend of King Albert I of Belgium. As such he was asked to give private lessons to the young prince Leopold III during World War I and in the years prior to his accession to the throne. This talented scientist was also a skilled pianist and blessed with a good tenor voice.

Biography

Buttgenbach, H. (1939) *Annales de la Société géologique de Belgique*, **62**, B285-291.
Buttgenbach, H. (1942) *Annuaire de l'Académie royale de Belgique*, **CVIII**, 35-70.
Buttgenbach, H. (1959) *Biographie nationale*, **30**, col. 275-279.
Mélon, J. & Donnay, J.D.H. (1980) Florilège des Sciences en Belgique, **II**, 437-450. Bruxelles: Académie royale de Belgique, Classe des Sciences.

Bibliography

Buttgenbach, H. (1942) *Annuaire de l'Académie royale de Belgique*, **CVIII**, 71-98.

COMBLAINITE

Gordon Comblain (1920-1996).

$Ni_6(Co^{3+})_2CO_3(OH)_{16}.4H_2O$

Trigonal.

Shinkolobwe, Haut-Katanga Province, Democratic Republic of the Congo.

Belgium: Royal Museum for Central Africa, Tervuren - Type: HT: RGM 2224.
USA: National Museum of Natural History, Washington - Type: T: 147188.

IMA N°: 1978-009

Piret, P. & Deliens, M. (1980) La comblainite, $(Ni^{2+}{}_x,Co^{3+}{}_{1-x})(OH)_2(CO_3)_{(1-x)/2}.yH_2O$, nouveau minéral du groupe de la pyroaurite. *Bulletin de Minéralogie*, **103**, 113-117.

http://www.handbookofmineralogy.org/pdfs/comblainite.pdf

Comblainite as a blue green clay between yellow orange becquerelite crystals.
Shinkolobwe, Haut-Katanga Province, D.R. Congo.
Royal Museum for Central Africa. Holotype (N° 2224).

Turquoise comblainite.
Shinkolobwe, Haut-Katanga Province, D.R. Congo.
F. Coune.

Gordon Comblain was born in Hackney (London) on August 11, 1920. He received a technical education in Saint-Gilles (Brussels). In 1946 after his military service (1939-1940) and following a long period of captivity, deportation and forced labour at the end of the Second World War, he attended an in-depth course organized by the *Société Internationale Forestière et Minière du Congo*, otherwise known as Forminière. In 1947, Comblain was offered a long-term contract by the *Société des Mines d'Etain du Ruanda-Urundi* (MINETAIN). As such, from March 17, 1947 until June 16, 1965, he was in charge of several mining and exploration missions in central Africa, and in particular in present-day Rwanda and Burundi. Among his major tasks were responsabilities in the field of gold, tin, tantalum and tungsten prospection, exploitation of mining plants, and supervision of the mining zone of Katumba, in Rwanda.

In August 1965, Comblain began working as a technician at the department of mineralogy of the Royal Museum for Central Africa (RMCA), in Tervuren. Here, under the stimulating guidance of Michel Deliens (*see*: deliensite), he became responsible for (a) the reorganization and maintenance of the mineralogical collections and the laboratory of X-ray analysis, (b) the purchase of new mineral specimens within Belgium and abroad, (c) the exchange of minerals with private mineral collectors and museums worldwide, and (d) contacts and scientific co-operation with university students preparing examinations and doing ongoing research. Comblain retired from the RMCA, in Tervuren, on September 1, 1985. He died in Brussels in 1996. Though he was born in England, Comblain was a Belgian citizen.

CORNETITE

Jules Cornet (1865-1929).

$Cu_3PO_4(OH)_3$

Orthorhombic.

Star of the Congo Mine (l'Etoile du Congo Mine, Kulukulu Mine), Lubumbashi, Haut-Katanga Province, Democratic Republic of the Congo.

Belgium: Laboratory of Mineralogy, University of Liège - Type: HT: 9130 to 9133.

IMA: G

Cesàro, G. (1912) Sur un nouveau minéral du Katanga. *Annales de la Société géologique de Belgique,* **39**, B241-242.

Schoep, A. (1927) Kristallen van cornetiet en hunne brekingsindices. *Natuurwetenschappelijk Tijdschrift,* **9** , 125-128.

Ungemach, H. (1928-1929) Précisions cristallographiques sur quelques minéraux du Congo belge. *Annales de la Société géologique de Belgique,* **52**, C75-85.

http://www.handbookofmineralogy.org/pdfs/cornetite.pdf

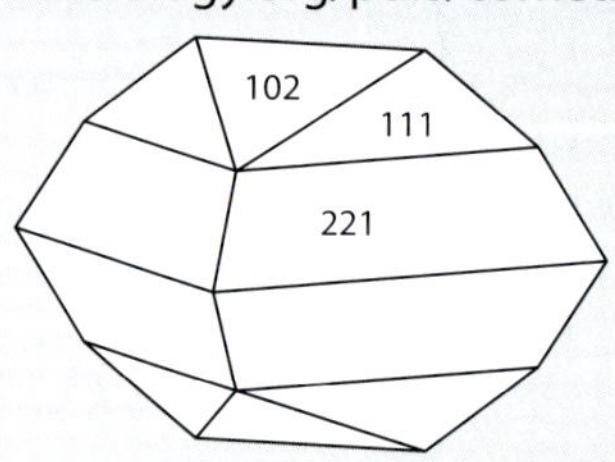

after Cesàro (1912)

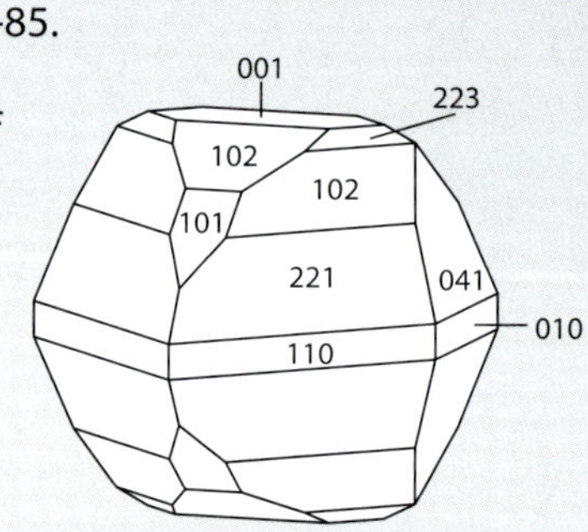

after Ungemach (1928-1929)

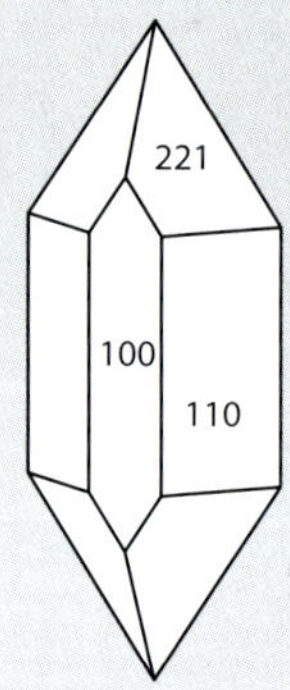

after Schoep (1927)

Small spherical aggregates of pale blue cornetite on a dark blue cornetite crust (2 generations). Lubumbashi, Haut-Katanga Province, D.R. Congo. P. De Bondt.

Intergrown cornetite crystals on chrysocolla.
Lubumbashi, Haut-Katanga Province, D.R. Congo.
J. Lhoest.

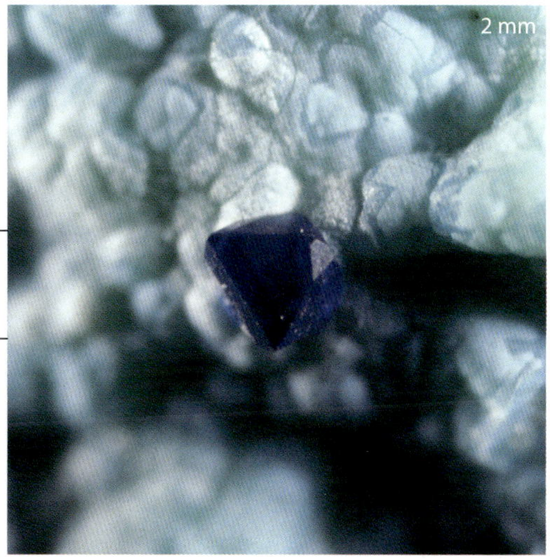

Double terminated short-prismatic cornetite (pseudo-octahedral).
Lubumbashi, Haut-Katanga Province, D.R. Congo.
P. De Bondt.

Prismatic cornetite crystals.
Lubumbashi, Haut-Katanga Province, D.R. Congo.
P. De Bondt.

Starlike aggregate of cornetite.
Lubumbashi, Haut-Katanga Province,
D.R. Congo.
Z. Gabelica.

Lath-shaped cornetite crystals.
Lubumbashi, Haut-Katanga Province, D.R. Congo.
Z. Gabelica.

Aggregate of long cornetite crystals.
Lubumbashi, Haut-Katanga Province, D.R. Congo.
J. Lhoest.

Jules Cornet was born on March 4, 1865, in Saint-Vaast (La Louvière), in the Hainaut Province. He was the son of a noted mining engineer, François-Léopold Cornet. Jules spent the greater part of his childhood in Cuesmes (Mons), where his father held an executive function at the *Charbonnage du Levant-du-Flénu*. As a schoolboy, he often accompanied his father in the field and as such got his first introduction to geology and field observations. In 1883, Jules enrolled at Ghent University (GU) for studies in medical sciences. After successfully finishing his first year studies in natural sciences, he accepted to become *préparateur* to the Institute of Zoology and Comparative Anatomy, headed by Felix Plateau. Under the stimulating influence of Alphonse Renard (*see*: renardite), Jules soon switched to geology and in October 1890, obtained his doctor's degree in mineral sciences with a dissertation dealing with phosphate-bearing rocks.

On May 18, 1891, Cornet left for the Congo Free State (CFS), where he participated in a scientific mission headed by Lucien Bia (*see*: *bialite*). This mission was called into existence to explore and safeguard the ore-bearing territories of Katanga against meddlesome authorities and nations. The observations he made in Katanga were of such an invaluable scientific interest that they won him national and international recognition, and years later earned him the reputation of being the founder of the geology of the present-day Democratic Republic of the Congo. The discovery of new copper deposits and the first evaluation by a scientist of their volumetric importance were among the major findings of the expedition. After the expedition had returned to Belgium (1893), Cornet continued to work as a *préparateur* at GU and participated at various occasions in geological missions and field trips in Northern and Eastern Europe, Russia and the

Mediterranean region. In August 1895, he went back to the CFS. At the request of the *Compagnie du Chemin de Fer du Congo* he now studied the proposed railway route between Matadi and Stanley-Pool, in the Lower Congo. By the end of the same year, he resumed his tasks at GU and proceeded with publishing the results of his pioneering research in Katanga and the Lower Congo.

In 1897 Cornet was offered a position as professor in mineralogy, geology and palaeontology at the *Ecole des Mines* in Mons. From then on his research interests centered on the geology of the Haine Basin (Hainaut Province), and he never returned to central Africa. When Renard passed away, Cornet also accepted academic responsabilities at GU. He was appointed associate professor and for many years lectured physical geography at the Faculty of Sciences, as well as history of commodities at the Special School of Commerce attached to the Faculty of Law. Between 1923 and 1926, in preparation of the changeover from French to Dutch at GU, Cornet, who didn't speak Dutch, was gradually relieved of his academic duties. He died in Mons on May 17, 1929, aged 64.

Biography

Anonymous (1935) Jules Cornet 1865-1929. Fondateur de la Géologie du Congo, Professeur à l'Ecole des Mines de Mons. Hommage des Disciples, des Anciens Elèves et des Admirateurs du Savant Géologue. 107 p. Mons: Associations des Ingénieurs de l'Ecole des Mines (Faculté Polytechnique).

Robert, M. (1954) *Annuaire de l'Académie royale de Belgique,* **CXX**, pp. 1-10.

Rijksuniversiteit te Gent. Liber memorialis 1913-1960, **IV**, 80-82, 1960. Gent: Uitgave van het Rectoraat.

Anonymous (1965) Jules Cornet 1865-1929. Centenaire de sa naissance 1965, 55-78. Bruxelles: Robert Louis.

Marlière, R. (1967) Florilège des Sciences en Belgique pendant le XIXe siècle et le début du XXe, 453-469. Bruxelles: Académie royale de Belgique, Classe des Sciences.

Bibliography

Robert, M. (1954) *Annuaire de l'Académie royale de Belgique,* **CXX**, 11-61.

Rijksuniversiteit te Gent. Liber memorialis 1913-1960, **IV**, 83-97, 1960. Gent: Uitgave van het Rectoraat.

Anonymous (1965) Jules Cornet 1865-1929. Centenaire de sa naissance 1965, 79-99. Bruxelles: Robert Louis.

COUSINITE

Jules Cousin (1884-1965).

$Mg(U^{4+})_2(MoO_4)_2(OH)_6.2H_2O$ (?)

Shinkolobwe, Haut-Katanga Province, Democratic Republic of the Congo.

IMA: Q

Vaes, J.F. (1958) Cousiniet, een nieuw uraanmineraal. *Geologie en Mijnbouw,* **20**, N°12, 449.
Fleischer, M. (1959) New Mineral Names. *The American Mineralogist,* **44**, 906-910.
Deliens, M. (1975) Une association de molybdates d'uranium de Shinkolobwe (région du Shaba, République du Zaïre). *Annales de la Société géologique de Belgique,* **98**, 155-160.
Deliens, M. & Piret, P. (1976) Nouvelles données sur une umohoite magnésienne de Shinkolobwe (région du Shaba, République du Zaïre). *Annales de la Société géologique de Belgique,* **99**, 205-209.

http://www.handbookofmineralogy.org/pdfs/cousinite.pdf

The description of cousinite as a new mineral by Vaes (1958) is very short and as such incomplete. Fleischer (1959) states that there are insufficient data and compares cousinite to moluranite. Deliens (1975) and Deliens & Piret (1976) suggest that cousinite could be a magnesium-bearing umohoite.

Descending from a family of Ardennese farmers, **Jules Cousin** was born on January 12, 1884 in Comblain-au-Pont, in the Liège Province. Following his primary and secondary schooling, he entered the University of Liège. Later on, he decided to continue his studies at the Catholic University of Leuven (CUL). Here he graduated as a mining engineer (1909) and an electrotechnical engineer (1910). On leaving the university, he served as a trainee with the *Union Minière du Haut-Katanga* (UMHK) and, as such, travelled to Elisabethville (modern Lubumbashi), in Katanga during the early days of 1911. In June of the same year, he was very actively involved in copper casting activities set up by the UMHK in Elisabethville. Between 1913 and 1917, Cousin was subsequently appointed as an assistant and director ad interim of UMHK. In 1917, he left Katanga for London and thereupon travelled to New York.

At the close of World War I, Cousin returned to Brussels. In 1919 he re-embarked for Katanga and shortly after his return to the former Belgian Congo he was appointed director general for Africa (1920-1922). There, in conjunction with Edgar Sengier (*see*: sengierite) and other colleagues, he constituted the so-called '*équipe Cousin*'. In 1923 Jules was called back to Brussels and appointed director of UMHK in Belgium and representative of the Board in Africa. In 1935, he became administrator and managing director in 1947. By the end of his professional career he served as administrator with residence in Africa, president of the *Comité de Direction* in Elisabethville and honorary president of UMHK (1962). Cousin passed away in Brussels on November 12, 1965.

Though Cousin's professional career was ruled by the rise and evolution of the copper industry in Katanga, he also achieved important humanitarian and social goals. He paid much attention to the social promotion of the native people, the education of the African women and the aid to students of modest financial means. One of his most important achievements was the foundation of the Saint Marguerite Institute for young African girls in Luishia, a locality situated north-west of present-day Lubumbashi. With this name he honoured his wife Marguerite Blaise, who died in 1944. On May 13, 1952, a Sengier-Cousin Museum was inaugurated in Likasi (Panda). Here almost all ore minerals discovered in Katanga over the years were on display.

Biography

Roger, E. (1967) *Biographie belge d'Outre-Mer*, **VI**, col. 241-246.

Van der Straeten, E. (1969) Jules Cousin, Pionnier, chef d'entreprise et homme de bien (1884-1965). *Académie royale des Sciences d'Outre-Mer, Classe des Sciences Morales et Politiques, nouvelle série*, **XXXVIII**-2, 39 p.

Kurgan-van Hentenrijk, G., Jaumain, S. & Montens, V. (eds) (1996) Dictionnaire des Patrons en Belgique. Les Hommes, les Entreprises, les Réseaux, 132-133. Bruxelles: De Boeck & Larcier, s.a. Dept. De Boeck Université.

DELHAYELITE

Fernand Delhaye (1880-1946).

$K_7Na_3Ca_5Al_2Si_{14}O_{38}F_4Cl_2$

Orthorhombic.

Mount Shaheru, Nyiragongo, Nord-Kivu Province, Democratic Republic of the Congo.

Belgium: Royal Museum for Central Africa, Tervuren - Type: CT: RGM 8037.

IMA: Approved in 1962 (without number).

Sahama, Th. G. & Hytönen, K. (1959) Delhayelite, a new silicate from the Belgian Congo. *Mineralogical Magazine*, **32**, 6-9.

Cannillo, E., Rossi, G. & Ungaretti, L. (1970) The crystal structure of delhayelite. *Rendiconti della Società Italiana di Mineralogia e Petrologia*, **26**, 63-75.

http://www.handbookofmineralogy.org/pdfs/delhayelite.pdf

Delhayelite was discovered by Sahama & Hytönen (1959) in a rock specimen from which earlier three new silicate minerals had already been described : götzenite, combeite and kirschsteinite.
Cannillo *et al.* (1970) established the crystal structure on the holotype provided by Thure Sahama.

Cleavage planes of delhayelite.
Khibiny Massif, Kola peninsula, Russian Federation.
Royal Belgian Institute of Natural Sciences.

Fernand Delhaye was born in Clermont-Ferrand, in central France, on February 4, 1880. His father was Belgian citizen. In 1897 Fernand moved to Mons with his parents, where he began his studies of mining engineer at the *Ecole des Mines du Hainaut* (Mons). Poor study results and an interruption due to military service resulted in him earning his academic degree no earlier than 1905 only. During 1906-1907 he continued his education in Mons and successfully concluded the programme of engineering geology with a dissertation on reef limestones of Frasnian age from the Dinant Basin. This may explain why from late 1907 until 1910, Delhaye was engaged by the *Société des Marbres de Merbes-le Château* (Hainaut Province).

Upon the advice of Jules Cornet (*see*: cornetite), Delhaye accepted to work in Katanga with the *Société Belge Industrielle et Minière du Katanga* (SIMKAT) for three years (1910-1913). Subsequently he did survey work in the Lower Congo (former Belgian Congo), and in what was then French Equatorial Africa. These activities were interrupted by the war. Delhaye spent the onset of World War I in London, and a draft notice mistakenly sent to his address in Mons did not reach him. This yielded a lot of trouble and unpleasant experiences once he got under arms. In September 1918, Delhaye and Maurice Sluys, a Belgian mining engineer who worked in the former Belgian Congo since 1912, left on a mission by government order for the former German East Africa. In the end they only reached Belgian Congo. There they spent one year and, through many field observations, managed to make a geological map of the Lower and Middle Congo. The following years (1921-1922) Delhaye headed a mission to the former Ruanda-Urundi, a United Nations trust territory under Belgian administration. Here he contrived to draft the geological map together with Achille Salée (*see*: saléeite).

In June 1922, Delhaye escaped death when, during field work in the Akanyaru Valley (former Ruanda-Urundi), a drunken native stabbed him and punctured one of his lungs with a dagger. From 1923 until 1938 he visited several African countries, serving the *Société Belge de Recherches Minières en Afrique* (REMINA) and other institutions. During this period he closely co-operated with Georges Borgniez (*see*: *borgniezite*) as well. Delhaye resided in Belgium during World War II and died quite unexpectedly at Saint-Gilles (Brussels) on December 15, 1946.

Biography

Robert, M. (1946) *Bulletin de la Société belge de Géologie, de Paléontologie et d'Hydrologie*, **55**, 232-235.
Passau, G. (1947) *Bulletin des Séances de l'Institut Royal Colonial Belge*, **18**, 169-172.
Anonymous (1965) Jules Cornet 1865-1929. Centenaire de sa naissance 1965, 109-114. Bruxelles: Robert Louis.
Sluys, M. (1967) *Biographie belge d'Outre-Mer*, **VI**, col. 285-300.

Bibliography

Passau, G. (1947) *Bulletin des Séances de l'Institut Royal Colonial Belge*, **18**, 173-176.
Anonymous (1965) Jules Cornet 1865-1929. Centenaire de sa naissance 1965, 114-119. Bruxelles: Robert Louis.

DELIENSITE

Michel Deliens (°1939).

$Fe^{2+}(UO_2)_2(SO_4)_2(OH)_2.3H_2O$

Orthorhombic.

Mas d'Alary, Le Bosc (Lodève), Hérault Department, France.

Belgium: Royal Belgian Institute of Natural Sciences, Brussels - Type: T: RC 4718.

IMA N°: 96-013

Vochten, R., Blaton, N. & Peeters, O. (1997) Deliensite, $Fe(UO_2)_2(SO_4)_2(OH)_2.3H_2O$, a new ferrous uranyl sulfate hydroxyl hydrate from Mas d'Alary, Lodève, Hérault, France. *The Canadian Mineralogist*, **35**, 1021-1025.

http://www.handbookofmineralogy.org/pdfs/deliensite.pdf

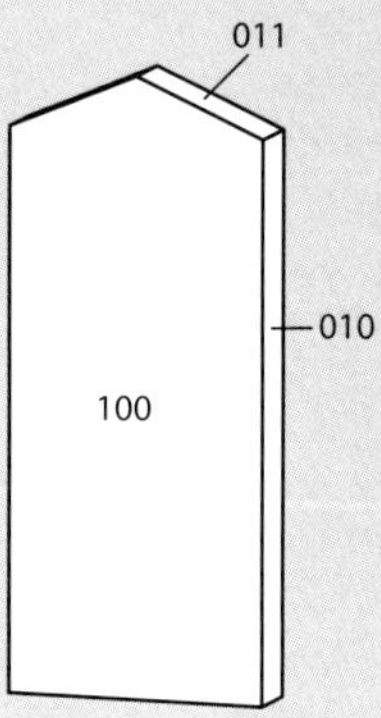

after Vochten *et al.* (1997)

SEM photograph of an aggregate of deliensite crystals.
Mas d'Alary, Lodève, Hérault Department, France.
Royal Belgian Institute of Natural Sciences.

SEM photograph of deliensite crystals.
Mas d'Alary, Lodève, Hérault Department, France.
Royal Belgian Institute of Natural Sciences.

Aggregates of submillimetric tabular deliensite crystals.
Mas d'Alary, Lodève, Hérault Department, France.
J.-C. Leydet.

Aggregates of submillimetric tabular deliensite crystals.
Mas d'Alary, Lodève, Hérault Department, France.
J.-C. Leydet.

Aggregates of submillimetric tabular deliensite crystals.
Mas d'Alary, Lodève, Hérault Department, France.
J.-C. Leydet.

Michel Deliens was born in Etterbeek (Brussels) on February 27, 1939. He studied at the Free University of Brussels (BFU), where he took a licentiate degree in geological and mineralogical sciences (1961). After one-year of practical training in hydrology with the Geological Survey of Belgium (GSB) sponsored by the Belgian Institute for the Encouragment of the Scientific Research Overseas (BIBWOO-IBERSOM) in Brussels, and following his military service, he accepted a position at the Royal Museum for Central Africa (RMCA) in Tervuren. His main task was to supervise the theoretical and practical education of African assistant-geologists. Concomitantly, he continued his scientific research and in 1973 received his PhD degree in geology from the *Université catholique de Louvain* (UCL). His dissertation, prepared under the guidance of Jules Moreau (*see*: moreauite), dealt with hydrated cobalt oxides from southern Katanga. Deliens remained contracted by the RMCA until 1979, when he joined the established service of this institution. As such he advanced rapidly from assistant (1979) to the highest possible rank of a scientific career (*chef de travaux*) (1982).

In 1987 Deliens was appointed head of the department of mineralogy and petrography of the Royal Belgian Institute of Natural Sciences (RBINS) in Brussels. Two years later he was also responsible for the departments of prehistory and anthropology of the same institute. In 1994 he succeeded Moreau as Belgian representative to the Commission on New Minerals, Nomenclature and Classification of the International Mineralogical Association (IMA-CNMNC)). He retired from active service during the spring of 2004.

For many years the scientific work of Deliens focused on mineralizations of copper, cobalt, nickel and uranium from Katanga. Later, his field of research encompassed other regions of the world as well, including Portugal, Germany, Romania, France, and Canada. During his professional career Deliens developed a close, lasting and most fertile collaboration with Paul Piret (*see*: piretite) of the UCL. They published many articles in co-authorship. About thirty papers of his that were published between the late 1970s and the beginning of this century, either as first author or in co-authorship, concern new mineral species. The majority of these are secondary uranium minerals discovered in Katanga and in the Kobokobo pegmatite, in Kivu (Democratic Republic of the Congo). Ten of these minerals are described in the present book [blatonite, comblainite, françoisite-(Nd), lepersonnite-(Nd), metavanmeersscheite, moreauite, mundite, oswaldpeetersite, piretite,vanmeersscheite].

DEMESMAEKERITE

Gaston Demesmaeker (1911-?).

$Pb_2Cu_5(UO_2)_2(Se^{4+}O_3)_6(OH)_6.2H_2O$

Triclinic.

Musonoi, Kolwezi, Lualaba Province, Democratic Republic of the Congo.

France:
Université de Paris - Type: HT: 9.097.
Musée de Minéralogie, Ecole des Mines de Paris - Type: CT.
Muséum national d'Histoire naturelle, Paris - Type: CT: 18136.
United Kingdom: National History Museum, London - Type: CT: BM 1969,46.

IMA N°: 65-019

Cesbron, F., Bachet, B. & Oosterbosch, R. (1965) La demesmaekerite, sélénite hydraté d'uranium, cuivre et plomb. *Bulletin de la Société française de Minéralogie et de Cristallographie*, **88**, 422-425.

http://www.handbookofmineralogy.org/pdfs/demesmaekerite.pdf

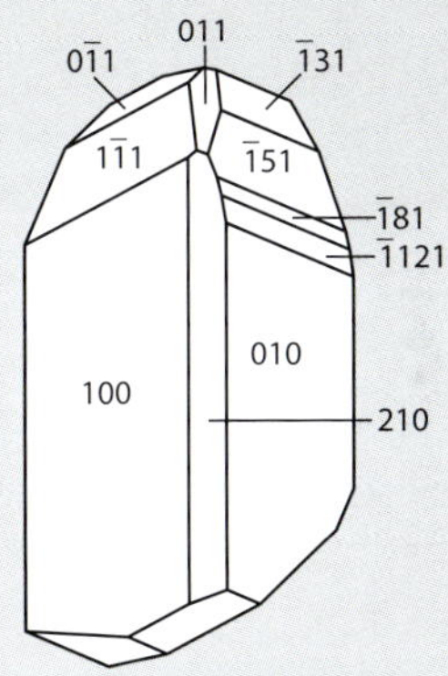

after Cesbron *et al.* (1965)

Khaki-coloured axe-shaped demesmaekerite crystals.
Musonoi, Lualaba Province, D.R. Congo.
P. De Bondt.

Dark green terminated demesmaekerite crystal.
Musonoi, Lualaba Province, D.R. Congo.
J.-C. Leydet.

Sheaf-like aggregate of khaki-coloured demesmaekerite.
Musonoi, Lualaba Province, D.R. Congo.
J. Lhoest.

Double terminated demesmaekerite crystals.
Musonoi, Lualaba Province, D.R. Congo.
F. Coune (N° 3056).

Dark green terminated demesmaekerite crystal.
Musonoi, Lualaba Province, D.R. Congo.
J. Lhoest.

Aggregate of greenish yellow demesmaekerite crystals. Musonoi, Lualaba Province, D.R. Congo. F. Coune N° 3039).

Radiated aggregate of khaki-coloured demesmaekerite crystals. Musonoi, Lualaba Province, D.R. Congo. P. De Bondt.

Gaston Demesmaeker was born in Halle, in the Flemish Brabant Province, on March 5, 1911. Following his primary and secondary education, he enrolled at the university. He graduated as an engineering geologist and took up his military service in July 1934. After serving the Belgian army he accepted to work for *Union Minière du Haut-Katanga* (UMHK) in the former Belgian Congo. In May 1940 he was drafted into the army again and during the war he participated in many missions on the African continent. Meanwhile he often stayed in Elisabethville (present-day Lubumbashi). Being incorporated into the Belgian Expeditionary Forces, he spent e.g. several months in Nigeria (September 1942 until mid-March 1943) and the Middle East (mid-March 1943 to late December 1943). At the end of the Second World War Demesmaeker was discharged. He continued his work for the UMHK in Katanga and was promoted to director of the geological department.

Gaston Demesmaeker is well-known for his research work on the tectonics of the stratiform copper deposits of Katanga, including the one of Shinkolobwe. This research was carried out in close collaboration with Armand François [*see*: françoisite-(Nd)] and Robert Oosterbosch (*see*: oosterboschite).

DERRIKSITE

Joseph Derriks (1912-1992).

$Cu_4(UO_2)(Se^{4+}O_3)_2(OH)_6$

Orthorhombic.

Musonoi, Kolwezi, Lualaba Province, Democratic Republic of the Congo.

France:
Musée de Minéralogie, Ecole des Mines de Paris - Type: CT: 9104 (label from Sorbonne).
Muséum national d'Histoire naturelle, Paris - Type: CT: 18135.
Université de Paris - Type: HT: 204 and 12251.

IMA N°: 71-033

Cesbron, F., Pierrot, R. & Verbeek, T. (1971) La derriksite, $Cu_4(UO_2)(SeO_3)_2(OH)_6.H_2O$, une nouvelle espèce minérale. *Bulletin de la Société française de Minéralogie et de Cristallographie*, **94**, 534-537.
Warin, R. (2003) La derriksite, extrapolation morphologique. *AGAB Minibul*, **36**, 8-17.

http://www.handbookofmineralogy.org/pdfs/derriksite.pdf

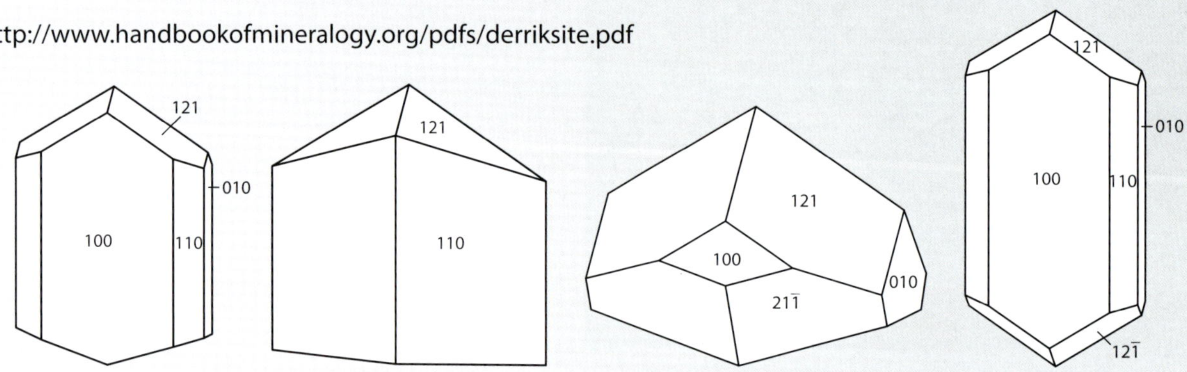

all drawings after Warin (2003)

Small aggregates of dark green derriksite crystals. Musonoi, Lualaba Province, D.R. Congo. Royal Museum for Central Africa (N° 10871).

Prismatic derriksite crystal with curved faces.
Musonoi, Lualaba Province, D.R. Congo.
J. Lhoest.

Well-developed prismatic derriksite crystals.
Musonoi, Lualaba Province, D.R. Congo.
J. Lhoest.

Aggregates of small, dark green derriksite crystals.
Musonoi, Lualaba Province, D.R. Congo.
J. Lhoest.

Rosette-like aggregates composed of flattened derriksite crystals.
Musonoi, Lualaba Province, D.R. Congo.
F. Coune (N° 1694).

Jean-Marie François Joseph Derriks (often referred to as **Joseph Derriks**) was born in Huy, in the Namur Province, on August 8, 1912. Following secondary schooling, he enrolled at the *Ecole des Mines* in Liège, where he attended the courses on mining engineering and engineering geology. By the mid-1930s he graduated from this institution. Before joining the *Union Minière du Haut-Katanga* (UMHK), he did his military service as a part of the engineer troops (1936-1937). In September 1938 he was admitted to the reserve officers' training corps of the Public Force for a ten year period. In May 1940 he was called under arms again. Belonging to the Belgian colonial troops in Africa (inclusive of the Belgian Expeditionary Forces), he spent a considerable part of World War II in the former Belgian Congo (in Elisabethville) and Nigeria.

For many years Derriks was heavily involved in the study of the Shinkolobwe site. He also paid much attention to other stratiform copper deposits in Katanga. In 1955 and 1958 he attended the United Nations International Conferences on the Peaceful Uses of Atomic Energy held at Geneva (Switzerland), where he presented important papers on the uranium of Katanga. Derriks was director of the UMHK for Shinkolobwe, and vice-president of the administrative council and managing director of the UM in Brussels. He retired from active service by the end of 1981. Being the representative of UMHK in Kinshasa in the 1960s, Derriks took an active part in the negotiations and discussions leading to the nationalization of the UMHK by the Congolese government by the end of 1966. He passed away in Angleur (Liège) on March 5, 1992.

Biography

Warin, R. (2003) Derriksite. *AGAB Minibul*, **36** (4), 80-89.

DEWINDTITE

Jan De Windt (1876-1898).

$H_2Pb_3(UO_2)_6O_4(PO_4)_4.12\ H_2O$

Orthorhombic.

Shinkolobwe, Haut-Katanga Province, Democratic Republic of the Congo.

France: Muséum national d'Histoire naturelle, Paris - Type: NT: 122.106.

IMA: G

Schoep, A. (1922) Sur la dewindtite, nouveau minéral radioactif. *Comptes rendus de l'Académie des Sciences, Paris,* **174**, 623-625.

Schoep, A. (1925) Nouvelles recherches sur la dewindtite: biréfringence, composition chimique et déshydratation de ce minéral. Comparaison avec la dumontite. *Bulletin de la Société française de Minéralogie*, **48**, 77-85.

Buttgenbach, H. (1947) Les minéraux de Belgique et du Congo Belge. 573 p., Paris: Dunod.

Piret, P., Piret-Meunier, J. & Deliens, M. (1990) Composition chimique et structure cristalline de la dewindtite, $Pb_3[H(UO_2)_3O_2(PO_4)_2]_2.12H_2O$. *European Journal of Mineralogy*, **2**, 399-405.

http://www.handbookofmineralogy.org/pdfs/dewindtite.pdf

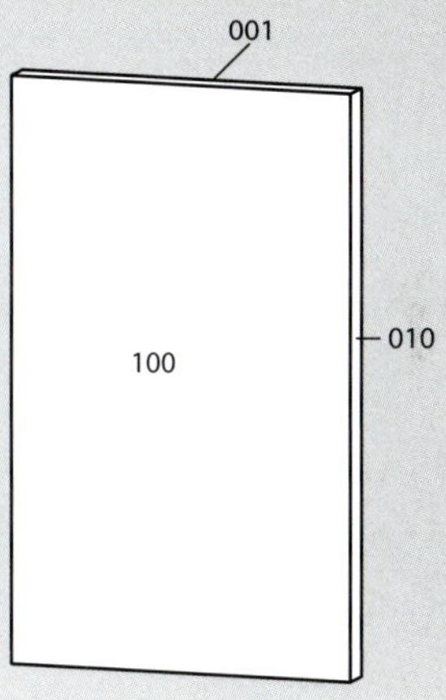

after Buttgenbach (1947)

See also ***renardite.***

Crust of small yellow dewindtite crystals with green metatorbernite. Shinkolobwe, Haut-Katanga Province, D.R. Congo. F. Coune (N° 3036).

Aggregates of dewindtite on top of lamellar metatorbernite.
Shinkolobwe, Haut-Katanga Province, D.R. Congo.
P. De Bondt.

Aggregates of small prismatic dewindtite crystals.
Shinkolobwe, Haut-Katanga Province, D.R. Congo.
F. Coune (N° 3036).

Thick lath-like dewindtite (renardite?) crystals.
La Faye Mine, Grury, Saône-et-Loire Department, F.
J.-C. Leydet.

Aggregates of radiating dewindtite (renardite?) crystals.
La Faye Mine, Grury, Saône-et-Loire Department, F.
E. Van Der Meersche.

Thin, lathlike, double terminated dewindtite (renardite?) crystals.
La Faye Mine, Grury, Saône-et-Loire Department, F.
F. Hubert (N° 1149).

Jan De Windt was born in Aalst, in the East-Flanders Province, on June 7, 1876. Merely sixteen years old, he entered the Faculty of Sciences of Ghent University (GU). At this university he was awarded a doctor's degree in natural sciences (July 1896). In reward of an innovative mineralogical and petrographic study on rocks from the massifs of Rocroi, Brabant and Stavelot, he was proclaimed laureate of the Inter-University Contest and invited to become an assistant to the laboratory of mineralogy and geology of GU, chaired by Alphonse Renard (*see*: renardite). To gain more field experience and theoretical scientific knowledge, De Windt travelled to Vienna and Berlin in 1896-1897. In Vienna, he worked under Albrecht Penck at the Geographic Institute of the local university. Before leaving Vienna, he spent several months at the Imperial Royal Natural History Court Museum of Austria-Hungary. At the Friedrich Wilhelm University of Berlin, De Windt attended classes given by Ferdinand Baron von Richthofen.

During his stay in Berlin, De Windt was asked to participate in an exploratory mission to Katanga. Despite unfavourable family-affairs, he accepted the proposal. This expedition, directed by lieutenant Charles Lemaire, was planned by the government of Congo Free State (CFS), received financial support by the *Comité Special du Katanga* (CSK) and took place under the official name of *Mission Scientifique du Katanga.* It lasted from 1898 to 1900. On April 16, 1898, De Windt and other members of the expedition left Amsterdam aboard S/S *Koning* with destination Chinde, near the mouth of the Zambezi River, in East Africa. Here, William Caisley, a gold prospector of British origin, joined the party. Upstream the Zambezi, the exploration team navigated the Shire River, before reaching Lake Nyassa and Lake Tanganyika. During the first days of August 1898, the forefront of the mission with Lt. Charles Lemaire and Léon Dardenne (the painter and draughtsman of the expedition), reached Moliro, a settlement at the southern extremity of Lake Tanganyika.

During the night of August 9 to August 10, a pirogue carrying De Windt, Caisley and eleven natives from Kituta to Moliro, was caught in a heavy storm and got lost with all hands on board. The lifeless bodies of both scientists were washed ashore a few hours after the terrible accident. Subsequently they were buried in the neighbourhood of Moliro, together with three natives who assisted them when the disaster occurred. With the death of De Windt, at the early age of 22, the geological community of GU lost one of its most promising scientific associates of the late 19th century. Though his professsional career hardly lasted two years, he authored four scientific papers dealing with very different research subjects.

Biography

Sabbe, H. (1898) *Tijdschrift van het Willems-fonds gewijd aan Letteren, Kunsten en Wetenschappen*. Derde Jaargang. Officieel gedeelte. CCXVI- CCXXI.

Le Mouvement géographique, 18 Septembre 1898, 15me année, n° 38, col. 470-471.

De Cock, A., Franckx, E., Mac Leod, J., Meert, H., Singelyn, J.L., Staes, G. & Van Branteghem, C. (1900) Almanak voor 1900 van het Fonds Jan De Windt gesticht tot ondersteuning der Vlaamsche Volksboekerijen, 10-13. Gent: I. Vanderpoorten.

Coosemans, M. (1952) *Biographie coloniale belge*, **III**, col. 928-930.

Bibliography

Coosemans, M. (1952) *Biographie coloniale belge*, **III**, col. 928-929.

Norbert Diderrich (1867-1925).

Shinkolobwe, Haut-Katanga Province, Democratic Republic of the Congo.

IMA: D

Vaes, J.F. (1947) Six nouveaux minéraux d'urane provenant de Shinkolobwe (Katanga): la billietite (urane hydraté de Ba), la vandendriesscheite (uranate hydraté de Pb), la masuyite (uranate hydraté de Pb), la richetite (uranate hydraté (?) de Pb), la studtite (carbonate hydraté d'urane) et la diderichite (carbonate hydraté d'urane). *Annales de la Société géologique de Belgique,* **70**, B212-225.

Frondel, C. & Meyrowitz, R. (1956) Studies on uranium minerals (XIX): Rutherfordine, diderichite and clarkeite. *The American Mineralogist,* **41**, 127-133.

Frondel & Meyrowitz (1956) pointed out that diderichite is identical with rutherfordite. The latter name has priority.

Norbert Diderrich was born in Vielsalm, in the Luxembourg Province, on April 26, 1867. Following the completion of his studies at the boarding school of the Brothers of the Christian Schools (Lasallian Brothers) in Carlsbourg (Paliseul), he enrolled at the Catholic University of Leuven (CUL). In 1889 he graduated as a mining engineer. At first, he worked in the quarries of Stoumont, in the Liège Province. Next, he joined the *Société nationale des chemins de fer vicinaux* (SNCV) and concurrently with this job he taught geology at the *Ecole agricole de l'Etat* in Carlsbourg. In 1890 he was recruited by the *Compagnie du Congo pour le Commerce et l'Industrie* (CCCI) in order to participate in an expedition under the command of Alexandre Delcommune. The goal of this mission was to explore Katanga and to try to convince Msiri, at the time controlling the Yeke Kingdom, to recognize the Belgian King Leopold II's sovereignty.

The Delcommune Expedition lasted from July 1890 until April 1893. During the tiring hiking tours through Katanga, Diderrich focused his scientific activities on the study of the subsoil and soil of the visited areas. Some time after the return of Delcommune and his companions to Belgium, the *Service agricole de l'Etat Indépendant du Congo* was established. In July 1894 Diderrich was appointed director of the registered office of the *Direction de l'Industrie et de l'Agriculture* in Boma. As such he was in charge of the inspection of the plantations of the Congo Free State (CFS) and had to set up the forestry services of Mayumbe. During the following years, ordered by the CFS, Diderrich travelled between Belgium and central Africa until he became director general of this organization. When the CFS was taken over by the Belgian government (in 1908) he was appointed member of the Colonial Council, a position he held until his death. Diderrich died in Brussels on April 23, 1925, at the age of 57.

Biography

Van den Abeele, M. (1952) *Biographie coloniale belge,* **III**, col. 239-244.

Joseph Donnay (1902-1994) and Gabrielle (Gai) Donnay-Hamburger.

$NaSr_3CaY(CO_3)_6.3H_2O$

Triclinic (pseudotrigonal).

Mont Saint-Hilaire, Montérégie region, Québec Province, Canada.

Canada:
Canadian Museum of Nature, Ottawa - Type: CT: 39394-39396.
Royal Ontario Museum, Toronto - Type: CT: M35222, M35544.
USA: National Museum of Natural History, Washington - Type: CT: 144522, 147191.

IMA N°: 78-007

Chao, G.Y., Mainwaring, P.R. & Baker, J. (1978) Donnayite, $NaCaSr_3Y(CO_3)_6.3H_2O$, a new mineral from Mont St. Hilaire, Quebec. *The Canadian Mineralogist*, **16**, 335-340.

http://www.handbookofmineralogy.org/pdfs/donnayitey.pdf

Donnayite-(Y) is often syntactically intergrown with ewaldite, mckelveyite and rarely synchisite. The differences in colours and patterns visible on the photographed specimens suggest intergrowths between donnayite-(Y) and the other minerals mentioned above.

Barrel-shaped donnayite-(Y) crystals with a pseudohexagonal contour.
Mount Saint-Hilaire, Québec, Canada.
Geonic collection.

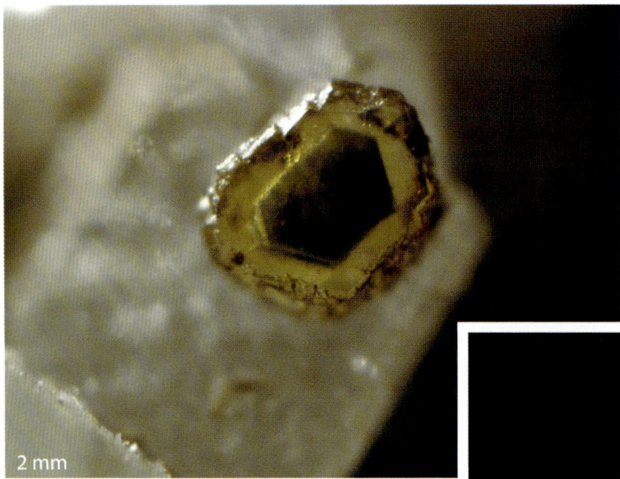

View parallel to the c-axis: pale yellow donnayite-(Y), enclosing ewaldite (?).
Mount Saint-Hilaire, Québec Province, Canada.
Royal Belgian Institute of Natural Sciences (N° C 4382).

Prismatic donnayite-(Y) crystals: difference in colour and surface in the left crystal suggests intergrowth.
Mount Saint-Hilaire, Québec Province, Canada.
Royal Belgian Institute of Natural Sciences (N° C 4382).

Aggregates of saucer-shaped pale yellow donnayite-(Y) crystals.
Mount Saint-Hilaire, Québec Province, Canada.
F. Hubert (N° 12819).

Joseph Donnay was born on June 6, 1902 in Grandville (Oreye), in the Liège Province. Following the completion of his early schooling in Waremme, Tongeren and Liège, he entered the University of Liège (ULg). Here he made the acquaintance of the mineralogists Giuseppe Cesàro (*see*: cesàrolite) and Henri Buttgenbach (*see*: buttgenbachite). In 1925 he graduated as a mining engineer. In the same year, thanks to a fellowship of the Commission for Relief in Belgium Educational Foundation, Donnay left Liège for Stanford University, California, USA. His stay at Stanford gave him the opportunity to investigate the copper deposit at the Engels Mine in Plumas County, California. This study earned him his PhD degree in 1929. After six months of prospecting for oil in Morocco, Donnay spent another year as research assistant at Stanford (1930-1931) before joining the Johns Hopkins University, Baltimore.

In 1939, Donnay was appointed professor to Laval University in Quebec, Canada. In 1944 he returned to the USA, where he served as a research chemist with the Hercules Powder Company. In 1945, Donnay returned to Johns Hopkins and became a professor. He spent the following year at the ULg before going back to Johns Hopkins. He retired in 1972, at the age of 70. After his official retirement from Johns Hopkins, Joseph and his wife Gabrielle (Gai) moved to Montreal, Canada, where Gabrielle was professor of crystallography at McGill University. At the age of 92, Joseph passed away on August 8, 1994 at the family house in Mont-Saint-Hilaire, Quebec, not far from the place where years before both donnayite and gaidonnayite were discovered. Gaidonnayite was named after Gabrielle (Gai) Donnay and was first described by Chao, G.Y. & Watkinson, D.H., 1974 (Gaidonnayite, $Na_2ZrSi_3O_9.2H_2O$, A new mineral from Mont St. Hilaire, Quebec, *The Canadian Mineralogist*, **12**, 316-319).

Joseph Donnay's studies on the relations between morphology and structure of crystals resulted in a new law of crystal morphology extending the law of Bravais. This law, formulated in 1937, was named after him and the American crystallographer David Harker and is generally known amongst mineralogists as 'the law of Donnay-Harker'. An extension or second generalization of the Donnay-Harker law was presented by Joseph and his wife in 1961. Donnay was a Belgian-American-Canadian citizen, whereas his wife was American-Canadian.

Biography

Donnay, J.D.H. (1962) For auld lang syne. *In*: Ewald, P.P. (ed) Fifty years of X-ray diffraction, 564-569.
Pabst, A. (1972) Presentation of the Roebling Medal of the Mineralogical Society of America for 1971 to J.D.H. Donnay. *The American Mineralogist*, **57,** 615-619.
Donnay, J.D.H. (1972) Acceptance of the Roebling Medal of the Mineralogical Society of America for 1971. *The American Mineralogist*, **57**, 620-625.
Perrault, G. & Stevenson, J.S. (1978) Preface of the Donnay Issue, *The Canadian Mineralogist*, **16**, 283.

Hubert Droogmans (1858-1938).

Shinkolobwe, Haut-Katanga Province, Democratic Republic of the Congo.

IMA: D

Buttgenbach, H. (1925) La droogmansite, nouvelle espèce minérale. *Annales de la Société géologique de Belgique*, **48**, B219-221.
Deliens, M. (1978) Droogmansite, a discredited species. *Bulletin de Minéralogie*, **101**, 561-562.

The description of *droogmansite* as a new mineral by Buttgenbach (1923) was very concise. New investigations were difficult because the specimen Buttgenbach used for his description seemed to be untraceable. Michel Deliens found a specimen that could correspond to that examined by Buttgenbach in the collection of the Laboratory of Mineralogy, University of Liège and proved that this mineral was kasolite in 1978.

Hubert Droogmans was born in Koersel-Stal (Beringen), in the Limburg Province, on September 6, 1858. Following his elementary and secondary schooling, he enrolled at the *Handelshogeschool* (High School for Commerce) in Antwerp, where he graduated in 1877. A few months later he was appointed chancellor to the consulate general of Belgium in Philadelphia, USA. It was the first step in a long and brilliant career in service of the Belgian King Leopold II, the Congo Free State (CFS) and the Belgian nation. During a stay in London (1885) Droogmans was invited to work at the finance department of the CFS. In 1894 he became secretary-general of this department. In this responsible position, he collaborated closely with Leopold II until the Belgian government took over the Congo in

1908. When the *Comité Spécial du Katanga* was founded (in 1900), Droogmans was designated chairman of this organization. He held this position for about thirty years. When Belgium took possession of its colony, Droogmans was appointed secretary-general to the newly established Ministry of Colonies and kept this high office until 1911. In 1929 he was invited to become vice-president of the *Section des Sciences Naturelles et Médicales de l'Institut Royal Colonial Belge* but he declined the invitation for health reasons. He died in Brussels on August 31, 1938.

Droogmans was much aware of the fact that in order to organize the economic and social development of Katanga, it was of utmost importance to have good connections to the Atlantic Ocean and to have access to reliable cartographic documents. Thanks to him, both goals were at least partly realized. The railway connection to the Atlantic Coast was realized thanks to his involvement. At his initiative and under his guidance the first map of Katanga (at the 1:1,000,000 scale) was published in 1904. This map is nowadays still known as '*carte Droogmans*'. Thanks to the support of both King Albert I of Belgium and the administrators of the *Compagnie du Katanga,* Droogmans was able to create a local Geographic and Geological Survey in the late 1910s. Higly stimulating the study of the geology, relief, hydrography, vegetation and superficial deposits of Katanga was merely one of his accomplishments. He was also very concerned with the public health in the former Belgian colony.

Biography

Robert, M. (1939) *Bulletin des Séances de l'Institut Royal Colonial Belge*, **X**, 46-50.

Robert, M. (1955) *Biographie coloniale belge,* **IV**, col. 242-247.

Bibliography

Robert, M. (1955) *Biographie coloniale belge,* **IV**, col. 247.

DUMONTITE

André Dumont (1809-1857).

$Pb_2(UO_2)_3(PO_4)_2O_2.5H_2O$

Monoclinic.

Shinkolobwe, Haut-Katanga Province, Democratic Republic of the Congo.

France: Muséum national d'Histoire naturelle, Paris - Type: CT: 125.11.

IMA: G

Schoep, A. (1924) La dumontite, nouveau minéral radioactif. *Comptes rendus de l'Académie des Sciences, Paris*, **179**, 693-695.

Thoreau, J., Van Meerssche, M. & Protas, J. (1958) Sur la dumontite de Shinkolobwe (Katanga). *Bulletin de la Société française de Minéralogie et de Cristallographie*, **81**, 63-65.

http://www.handbookofmineralogy.org/pdfs/dumontite.pdf

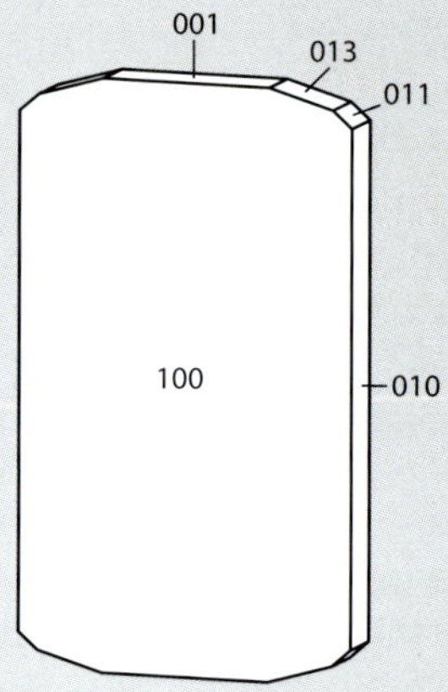

after Schoep (1924)

Crust of yellow dumontite and green metatorbernite. Shinkolobwe, Haut-Katanga Province, D.R. Congo. F. Coune (N° 3066).

Aggregates of lathlike yellow dumontite.
Shinkolobwe, Haut-Katanga Province, D.R. Congo.
F. Coune (N° 3066).

Crystal of yellow dumontite.
Shinkolobwe, Haut-Katanga Province, D.R. Congo.
Royal Belgian Institute of Natural Sciences (N° RC 3535).

André-Hubert **Dumont** was born in Liège on February 15, 1809. At the age of fifteen, his parents sent him to Paris to acquire commercial skills. As he showed little or no interest in commercial matters, he returned to his birthplace. There, he pursued studies of natural sciences. Under the influence of his father, who was a mine surveyor, he became wildly enthusiastic about mines and all things linked to it. In 1828 André Dumont was appointed surveyor of mines and at the age of 21 he was declared laureate of the Royal Academy of Belgium with a memoir on the geological constitution of the Liège Province (*see*: davreuxite). At about the same time he enrolled at the department of physical and mathematical sciences of the University of Liège (ULg). He obtained his doctor's degree in sciences in 1835. A few months later, he was appointed professor of mineralogy and geology to the Liège alma mater.

In 1836 the Belgian government asked him to prepare a geological map of the Liège, Hainaut and Luxembourg Provinces and about one year later and at his explicit request this assignment was extended in order to cover the entire Belgian territory. At the expense of great physical efforts and in spite of a heavy educational task at the university, Dumont managed to finish the mapping activities and to publish the ordered cartographic document in 1853. The result was a nine-sheet geological map at the 1:160,000 scale, yielded by sixteen years of hard labour. This work was carried out by a single person who did not hesitate to travell about 100,000 km on foot and on horseback across the country and irrespective of weather conditions in order to achieve his goal. A few years earlier (1849), Dumont had already published a geological map of Belgium and the neighbouring regions at the scale of 1:800,000.

Little by little, it became apparent that the many years of field work had considerably affected Dumont's health. Upon the advice of his physician and his family he agreed to take some months of rest abroad, in a region with mild climatological conditions. By the end of 1853 Dumont returned to Belgium after a tiring travel through Europe and Turkey. Instead of thoroughly enjoying life and recuperating of the loss of physical strength, he took advantage of his journey abroad to do geological observations and to participate in many excursions with colleagues from different countries. The scientific data he collected during that period was used to prepare a geological map of Europe at the 1:4,000,000 scale. It was displayed at the World Exhibition of Paris in 1855. Shortly afterwards, in spite of failing health, Dumont accepted to be assigned as rector of the ULg. He died from a brain haemorrhage in Liège on February 28, 1857, at the early age of 48.

Biography

d'Omalius d'Halloy, J.-B. (1858) *Annuaire de l'Académie royale des Sciences, des Lettres et des Beaux-Arts de Belgique,* **24**, 91-100.

Fayn, J. (1862) *Revue universelle des Mines, de la Métallurgie, des Travaux publics, des Sciences et des Arts appliqués à l'Industrie*, **15**, 1-96.

Fayn, J. (1864) André Dumont, sa vie et ses travaux. 275 p. Paris and Liège: Noblet et Baudry.

Liber memorialis. L'Université de Liège depuis sa fondation, col. 216-249, 1869. Liège: J.-G. Carmanne.

Dewalque, G. (1878) *Biographie nationale*, **6**, col. 283-295.

Thoreau, J. (1967) Florilège des Sciences en Belgique pendant le XIXe siècle et le début du XXe, 439-451. Bruxelles: Académie Royale de Belgique, Classe des Sciences.

Bibliography

Liber memorialis. L'Université de Liège depuis sa fondation, col. 250-255, 1869. Liège: J.-G. Carmanne.

EYLETTERSITE

Mrs. Lea Van Wambeke, maiden name: Eyletters

$Th_{0.75}Al_3(PO_4)_2(OH)_6$

Trigonal.

Kobokobo, Sud-Kivu Province, Democratic Republic of the Congo.

Belgium: Private collection of Leopold Van Wambeke.

IMA N°: 69-035

Van Wambeke, L. (1972) Eylettersite, un nouveau phosphate de thorium appartenant à la série de la crandallite. *Bulletin de la Société française de Minéralogie et de Cristallographie*, **95**, 98-105.

http://www.handbookofmineralogy.org/pdfs/eylettersite.pdf

Cream-coloured eylettersite.
Kobokobo, Sud-Kivu Province, D.R. Congo.
L. Vandenberghe.

Cream-coloured eylettersite.
Kobokobo, Sud-Kivu Province, D.R. Congo.
H. Dillen (7.2.18.10/1).

FOURMARIERITE

Paul Fourmarier (1877-1970).

$Pb_{1-x}O_{3-2x}(UO_2)_4(OH)_{4+2x}.4H_2O$

Orthorhombic.

Shinkolobwe, Haut-Katanga Province, Democratic Republic of the Congo.

France: Muséum national d'Histoire naturelle, Paris - Type: CT: 124-181.
Belgium: Laboratory of Mineralogy, University of Liège - Type: HT: 16.871 and 16.872.

IMA: G

Buttgenbach, H. (1923-1924) Minéralogie du Congo belge. La fourmariérite, nouvelle espèce minérale. *Annales de la Société géologique de Belgique*, **47**, C41-43.
Palache, C., Berman, H. & Frondel C. (1944) The system of mineralogy. Volume I: elements, sulfides, sulfosalts, oxides. 7th Edition, 834 p., John Wiley & Sons Inc.

http://www.handbookofmineralogy.org/pdfs/fourmarierite.pdf

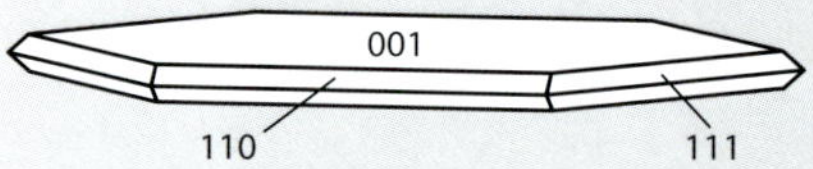

after Palache *et al.* (1944)

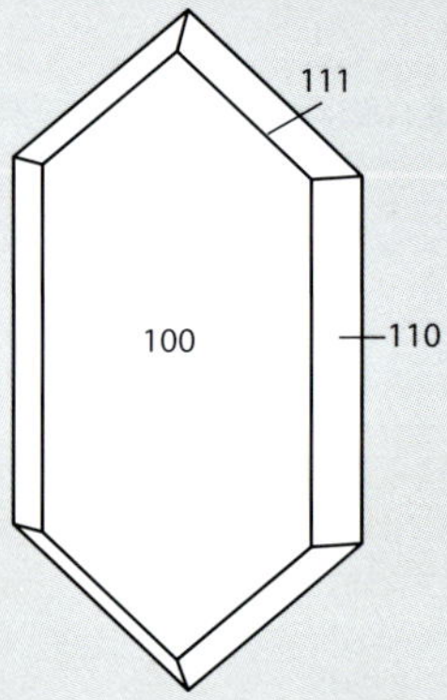

after Buttgenbach (1923-1924)

Spherical aggregates of platy orange brown fourmarierite crystals. Shinkolobwe, Haut-Katanga Province, D.R. Congo. F. Hubert (N° 90726).

Aggregates of tabular brown yellow fourmarierite crystals. Shinkolobwe, Haut-Katanga Province, D.R. Congo. F. Coune (N° 3067).

FOURMARIERITE

Short-prismatic reddish fourmarierite crystals.
Shinkolobwe, Haut-Katanga Province, D.R. Congo.
Royal Belgian Institute of Natural Sciences (N° 2827).

Tabular orange red fourmarierite crystal.
Shinkolobwe, Haut-Katanga Province, D.R. Congo.
F. Hubert (N° 90726).

Tabular reddish brown fourmarierite crystal.
Shinkolobwe, Haut-Katanga Province, D.R. Congo.
R. van Dooren.

Prismatic brownish red fourmarierite crystal.
Shinkolobwe, Haut-Katanga Province, D.R. Congo.
F. Coune (N° 3004).

Paul Fourmarier was born in La Hulpe, in the Walloon Brabant Province, on Christmas Day, 1877. He received his candidate's education of engineer at the Free University of Brussels (BFU) but finished his studies of civil mining engineer at the University of Liège (ULg) in 1899. The following year he entered the *Corps des Mines* and obtained his degree in engineering geology in Liège in 1901. He was the first student in this specialization since a mining school was established in this city (1900). The first years of Fourmarier's professional career were devoted to stratigraphical, structural and lithological studies of the Wallonian and the Campine coal-basins, including those in the neighbourhood of Liège, Theux, Andenne, the Herve region, the Hainaut and Belgian Limburg. These years were determinative for his scientific work during his entire academic career and even until his death. During this time, he also initiated his observations on the structural evolution and significance of the tectonic window of Theux. This subject would occupy his mind and research activities during his entire life.

By request of Max Lohest (*see*: *lohestite*), Fourmarier was appointed assistant to the Liège alma mater in 1901. From 1907 to 1920 he served as a tutor and in 1920 - the year he left the *Corps des Mines* - he was appointed full professor to the Faculty of Sciences and the Technical Faculty of this institution. Fourmarier's teaching programme included applied geology, hydrogeology and geography of mines and industry. With the death of Lohest (1927) his lecturing responsabilities expanded and from then on also included general geology and physical geography. Fourmarier became emeritus professor in 1948 but continued to work on geological problems and topics up to his death. He passed away in Liège on January 20, 1970, at the advanced age of 92.

Fourmarier acquired a very solid reputation with his research on the structure of the tectonic window of Theux and the Ardennes in general. He wrote numerous papers on the schistosity, a geologic feature that held his attention for over half a century. He published several outstanding synthetic works and memoirs (*'Eléments de Géologie'*, *'Principes de Géologie'*, and *'Vue d'ensemble de la Géologie de la Belgique. Ses enseignements dans le domaine de la géologie générale'*). His contribution to the knowledge of the geology of the former Belgian Congo and other African countries is far from being negligible. Fourmarier's bibliography is an enumeration of nearly 600 treatises, scientific papers and notes, and all kinds of reports.

Biography

Michot, P. (1970) *Annales de la Société géologique de Belgique*, **93**, 425-429.

Denaeyer, M.-E. (1971) *Bulletin des Séances de l'Académie royale des Sciences d'Outre-Mer*, **17**, 71-81.

Calembert, L. (1971) *Bulletin de la Société géologique de France*, 7me série, **13,** 205-210.

Calembert, L. (1989) *Biographie belge d'Outre-Mer*, **VII-C**, col. 141-149.

Delmer, A. (2003) *Annuaire de l'Académie royale de Belgique*, **CLXIX**, 3-9.

Bibliography

Denaeyer, M.-E. (1971) *Bulletin des Séances de l'Académie royale des Sciences d'Outre-Mer,* **17**, 81-85 (papers with regard to Africa).

Calembert, L. (1971) *Bulletin de la Société géologique de France*, 7me série, **13**, 210-218.

Calembert, L. (1989) *Biographie belge d'Outre-Mer*, **VII-C**, col. 149-153 (selected papers).

Delmer, A. (2003) *Annuaire de l'Académie royale de Belgique,* **CLXIX**, 10-38.

FRANÇOISITE-(Nd)

Armand François (° 1922).

$Nd(UO_2)_3(PO_4)_2O(OH).6H_2O$

Monoclinic.

Kamoto-E, Kolwezi, Lualaba Province, Democratic Republic of the Congo.

Belgium: Royal Belgian Institute of Natural Sciences, Brussels - Type: HT: RC 3.512; RN 6380.

IMA N°: 87-041

Piret, P., Deliens, M. & Piret-Meunier, J. (1988) La françoisite-(Nd), nouveau phosphate d'uranyle et de terres rares; propriétés et structure crystalline. *Bulletin de Minéralogie,* **111**, 443-449.

http://www.handbookofmineralogy.org/pdfs/francoisitend.pdf

FRANÇOISITE-(Ce)

Ce-dominant analogue of françoisite-(Nd)

$Ce(UO_2)_3(PO_4)_2O(OH).6H_2O$

Monoclinic.

Mt. Painter N°2 workings, Radium Ridge, North Flinders Ranges, Australia.
La Creusaz uranium mine, Les Marécottes, Trient Valley, Valais, Switzerland.

Switzerland:
Geological Museum of Lausanne - Type : CT 79288.
Geological Museum of Lausanne - Type : HT 58321.

IMA N°: 2004-029

Meisser, N., Brugger, J., Ansermet, S., Thélin, P. & Bussy, F.. Françoisite-(Ce), (Ce, Nd, Ca)$[(UO_2)_{30}(OH)(PO_4)_2]$ $(H_2O)_6$, a new mineral from the Aiguilles Rouges massif, western Swiss Alps. *European Journal of Mineralogy*, Publication pending.

Roth, Ph. (2007) Minerals first discovered in Switzerland and minerals named after Swiss individuals. 239 p., Zurich: Philippe Roth.

Crust of platy brownish yellow françoisite-(Nd) crystals. Kamoto-East Mine, Kolwezi, Haut-Katanga Province, D.R. Congo. Royal Belgian Institute of Natural Sciences. Holotype (RN 6380).

Yellow françoisite-(Nd) crystals with green metatorbernite. Kamoto-East Mine, Kolwezi, Haut-Katanga Province, D.R. Congo. F. Coune.

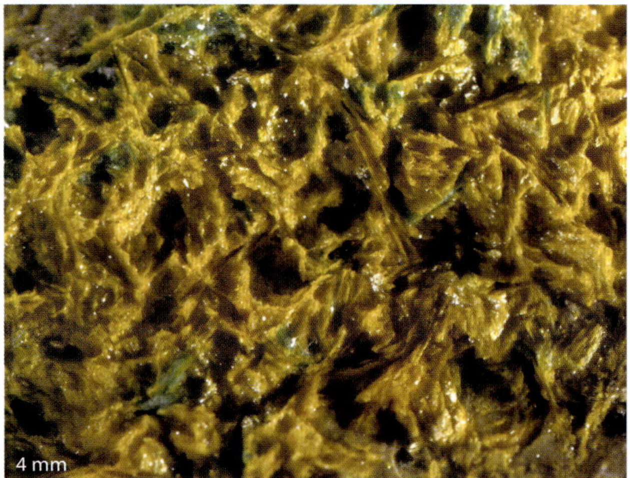

Reticular crust of brownish yellow françoisite-(Nd) crystals. Kamoto-East Mine, Kolwezi, Haut-Katanga Province, D.R. Congo. J. Lhoest.

Crust of brownish yellow françoisite-(Nd) crystals with green metatorbernite. Kamoto-East Mine, Kolwezi, Haut-Katanga Province, D.R. Congo. P. De Bondt.

Armand Pierre François was born on December 28, 1922, in Aljustrel (Portugal), where his father was employed as an electro-technical engineer in a pyrite mine. At the age of 4, Armand and his parents moved to Slata (Tunisia), a village in an ore-rich region near the Algerian border. He spent the greater part of his secondary education at the College of the Marists in Tunis (1933-1938). Yet the last year of this secondary schooling was in Liège, followed by an additional year in preparation for mining engineering studies. In 1940, Armand entered the University of Liège (ULg) and pursued studies in mining engineering until 1944. At the close of the fourth year of his university study (1944), he joined the Belgian army as a volunteer. As such he was sent to Northern Ireland at the close of World War II. Following this interruption, he returned to the university and graduated as a mining engineer in 1946.

After his graduation François accepted to work at the coal-mine of Wérister, in the Liège Province. In 1947, *Union Minière du Haut-Katanga* (UMHK) was looking for an engineer geologist willing to work in Katanga. To this end, the company was prepared to pay the studies of a young mining engineer interested in serving as an engineer geologist. As François was interested in both the proposal and in a professional career in the former Belgian Congo, he took up the post. In December 1948, he graduated as an engineering geologist from the ULg and left for Katanga two months later. From 1949 until 1977, François worked as a geologist in Katanga, first for the UMHK and later for the Gécamines (*Générale des Carrières et des Mines*). During this period, his field studies and active prospecting activities focused on the Tenke-Fungurume area, and on the copper and cobalt mineralizations occurring at Kolwezi. He was appointed director of the department of geology in Likasi in 1965, a position he held until 1978. Subsequently, he retired from Gécamines, and returned to Belgium where he was appointed director to the department of geology of UM (the former UMHK) in Brussels. He retired in October 1984. After retirement François continued to put his expertise at the disposal of mining societies and doing so made many journeys to England, Canada, and South Africa. He also returned to Katanga several times.

FRANSOLETITE

André-Mathieu Fransolet (° 1947).

$Ca_3Be_2(PO_4)_2(PO_3OH)_2.4H_2O$

Monoclinic.

Tip Top pegmatite, Custer, Custer County, South Dakota, USA.

USA: National Museum of Natural History, Washington - Type: HT: 149607 (donated by John Sinkankas, 1982).

IMA N°: 82-096

Peacor, D.R., Dunn, P.J., Roberts, W.L., Campbell, T.J. & Newbury, D. (1983): Fransoletite, a new calcium beryllium phosphate from the Tip Top pegmatite, Custer, South Dakota. *Bulletin de Minéralogie,* **106**, 499-503.

http://www.handbookofmineralogy.org/pdfs/fransoletite.pdf

SEM photograph of an aggregate of fransoletite crystals. Tip Top Mine, Custer, Custer County, South Dakota, USA. Laboratory of Mineralogy, University of Liège.

Aggregates of whitish fransoletite crystals.
Tip Top Mine, Custer, Custer County,
South Dakota, USA.
A.-M. Fransolet.

Sheaf-like aggregate of colourless fransoletite crystals.
Tip Top Mine, Custer, Custer County, South Dakota, USA.
A.-M. Fransolet.

Curved aggregate of platy whitish fransoletite.
Tip Top Mine, Custer, Custer County, South Dakota, USA.
A.-M. Fransolet.

PARAFRANSOLETITE

$Ca_3Be_2(PO_4)_2(PO_3OH)_2.4H_2O$

Triclinic.

Tip Top pegmatite, Custer, Custer County, South Dakota, USA.

USA:
Los Angeles County Museum of Natural History, Los Angeles - Type: HT: 35696 (parts of a single specimen).
National Museum of Natural History, Washington - Type: HT: 168432.

IMA N°: 89-049

Kampf, A.R., Dunn, P.J. & Foord, E.E. (1992) Parafransoletite, a new dimorph of fransoletite from the Tip Top Pegmatite, Custer, South Dakota. *American Mineralogist*, **77**, 843-847.

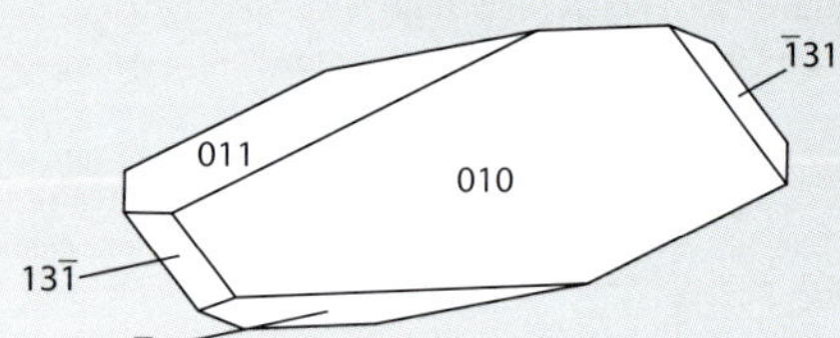

after Kampf *et al.* (1992)

http://www.handbookofmineralogy.org/pdfs/parafransoletite.pdf

Aggregate of micaceous whitish parafransoletite.
Tip Top Mine, Custer, Custer County, South Dakota, USA.
A.-M. Fransolet.

Star-like aggregate of micaceous whitish parafransoletite.
Tip Top Mine, Custer, Custer County, South Dakota, USA.
A.-M. Fransolet.

Born at Heusy (Verviers), in the Liège Province, on May 8, 1947**, André-Mathieu Fransolet** received his secondary education at the Saint-Michael Institute in Verviers. In 1965 he enrolled at the University of Liège (ULg) where he obtained his licentiate's degree in geological and mineralogical sciences in October 1969. From 1967 until 1969, he served as a student-assistant to the Institute of Mineralogy in Liège. Subsequently he was a research trainee to the Belgian National Fund for Scientific Research (NFWO-FNRS) for a short period of time and studied complements of general inorganic chemistry and physical chemistry between 1971 and 1972. In 1975 Fransolet graduated as a doctor in geological and mineralogical sciences at the ULg.

At the beginning of his professional career Fransolet served as an assistant to the Institute of Mineralogy of the ULg, a post he held from 1970 until October 1978. From November 1, 1978 to September 30, 1979, he was a research fellow of the Alexander von Humboldt Fondation and worked at the time under Werner Schreyer, professor of mineralogy at the Ruhr University in Bochum, Germany. Upon his return to Liège, in the fall of 1979, Fransolet became research associate of the NFWO-FNRS in Brussels. In 1990 he accepted the position of lecturer at the Faculty of Sciences of the ULg and on January 1, 1998, he was promoted to full professor at this institution. Up till now he teaches classes of mineralogy, crystallography, radiocrystallography and crystal-chemistry.

The past and present research activities of André-Mathieu Fransolet primarily concern crystal-chemistry, systematic mineralogy and mineralogical analysis of e.g. quartz veins from the Belgian Ardennes. This, nevertheless, does not interfere with his strong involvement in projects of applied mineralogy. The subjects of his applied research are very varied. Some of them are carried out on demand of national and international metallurgical concerns, whereas others dealt with mineralogical research on inorganic dust particles found in the lungs of hospital patients. As to fieldwork, he was primarily active in Africa, paying particular attention to Fe-Mn phosphate associations occurring in granite pegmatites from the Anti-Atlas (Morocco), southern Madagascar and the Gatumba region in Rwanda. Thanks to Fransolet and other mineralogists at work in other laboratories in Belgium or abroad, the minerals stavelotite-(Nd) and vantasselite, which are also illustrated in this book, were recognized as new mineral species.

JEDWABITE

Jacques Jedwab (°1925).

Fe_7Ta_3

Hexagonal.

Avrorinskii placer, Aktai River, Nizhniy-Tagil, Middle Urals, Russian Federation.

Russia:
Fersman Mineralogical Museum, Moscow - Type: T: 88656, 88704.
VGM-Moscow - Type: T.

IMA N°: 95-053

Novgorodova, M.I., Generalov, M.E. & Trubkin, N.V. (1997) Jedwabite $Fe_7(Ta,Nb)_3$ - a new mineral in association with tantal- and niobocarbides from Pt-bearing placers (in Russian with English abstract). *Zapiski Vserossiyskogo Mineralogischeskogo Obchshestvo,* **126/2**, 100-104.

Mandarino, J.A. (1998) Abstracts of new mineral descriptions. *The Mineralogical Record,* **29**, 467-479.

http://www.handbookofmineralogy.org/pdfs/jedwabite.pdf

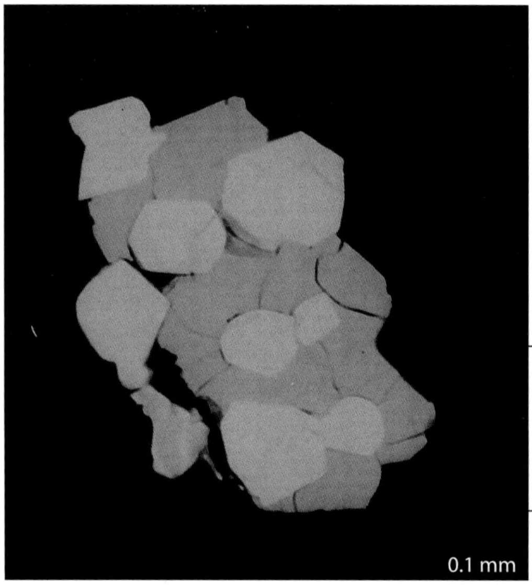

*SEM photograph (back-scattered mode) of jedwabite between niobocarbide crystals.
Avrorinskii placer, Aktai River, Nizhniy-Tagil, Middle Urals, Russian Federation.
Photograph courtesy of J.Jedwab .*

Jacques Jedwab was born in Charleroi, in the Hainaut Province, on May 7, 1925. Polish by birth, he chose Belgian citizenship in 1941. In 1945, at the completion of his primary schooling at Gilly, Charleroi and Saint-Gilles (Brussels) and following his secondary education at the Atheneum of Saint-Gilles he enrolled at the Free University of Brussels (BFU). In 1949 he obtained his licentiate degree in geological and mineralogical sciences. During the academic years 1949-1950 and 1952-1953 (a two year gap due to military service) he held a grant of the Institute for the Encouragement of Scientific Research in Industry and Agriculture (IWONL-IRSIA) in Brussels. At the end of his grant, he obtained his PhD with a dissertation on trace elements in pegmatites, prepared under the direction of Ivan de Magnée of the BFU.

From 1953 to 1955 Jedwab served as a post-doctoral research fellow of IWONL-IRSIA and from 1955 to 1958 as a researcher at the Center for Applied Geochemistry of the BFU. Within the context of this grant he developed methods for the geochemical prospection of some specific metals (W, U) and kimberlites. From 1958 to 1960, he was appointed research associate to the Inter-University Institute for Nuclear Sciences (IIKW-IISN) in Brussels and studied natural uranium-carbonaceous matter associations. During the following decade, Jedwab subsequently became associate lecturer (1960), lecturer (1963), extraordinary professor (1965) and full professor (1969) at the BFU. He was appointed emeritus professor in 1990.

The research projects carried out by Jedwab during his professional career mainly dealt with: (a) rocks and dust particles collected during manned NASA missions to the Moon, (b) carbonaceous meteorites and cosmic dust extracted from manganese nodules (3), (c) heavy metal minerals suspended in deep ocean waters, (d) rare minerals in hydrothermal chimneys and Red Sea troughs, (e) medical mineralogy (pneumology, urology), and (f) mineralogy of Platinum Group Elements (PGE) from various countries in North and South America, Europe and Africa, mostly after 1990, year of his retirement.

He continues to do scientific work, notably on Platinum Group Minerals (PGM), mainly from the former Belgian Congo (palladinite, luberoite, verbeekite), Brazil (ouro preto) and USA (PGE-bearing bitumens from Nevada).

Bibliography

http://www.ulb.ac.be/sciences/upgm/mguide

LEGRANDITE

Charles Legrand (1873-?).

$Zn_2AsO_4(OH).H_2O$

Monoclinic.

Flor de Peña Mine, Lampazos de Naranjo, Nueva León, Mexico.

USA:
HMM-Cambridge - Type: HT (part of): 92567 (donated by the Natural History Museum, London through Julien Drugman, 1933)
National Museum of Natural History, Washington - Type: T: 114810 (donated by Max Hey, 1960)
United Kingdom: National History Museum, London - Type: HT (part of): BM 1932,131.

IMA: G

Drugman, J. & Hey, M.H. (1932) Legrandite, a new zinc arsenate. *Mineralogical Magazine*, **23**, 175-178.

http://www.handbookofmineralogy.org/pdfs/legrandite.pdf

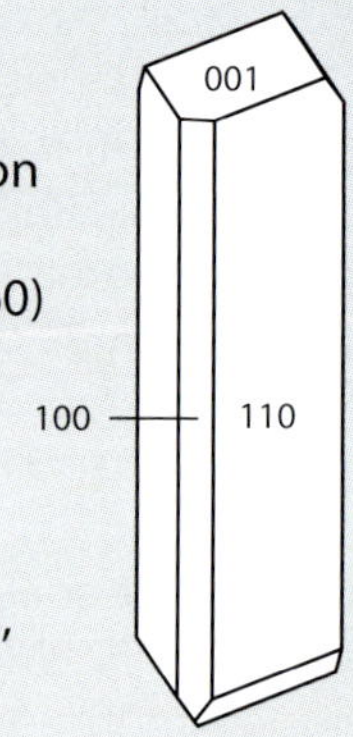

after Drugman & Hey (1932)

Aggregates of prismatic legrandite crystals.
Ojuela Mine, Mapimi, Durango Province, Mexico.
P. De Bondt.

Sheaf-like aggregate of brownish yellow legrandite crystals.
Ojuela Mine, Mapimi, Durango Province, Mexico.
P. De Bondt.

Double terminated prismatic legrandite crystal. Ojuela Mine, Mapimi, Durango Province, Mexico. F. Hubert (N° 10579).

Clear, double terminated prismatic legrandite crystal. Ojuela Mine, Mapimi, Durango Province, Mexico. F. Hubert (N° 1344).

According to Peter Cristofono (personal communication to Ernst A.J. Burke, former chairman of the IMA-CNMNC) **Charles Legrand** was born in 1873 in Liège. After receiving his primary and secondary education in Liège he enrolled at the *Ecole Industrielle* of his birthplace, attending classes in mechanical and electrical engineering (1887-1891). He subsequently graduated with a degree of bachelor in applied sciences from the University of Toronto, Canada.

Between 1897 and 1908, Legrand served several mining companies in Canada and the USA, and worked at the Faculty of Applied Science, University of Toronto for a certain period of time. In the early 20th century, Legrand also stayed in Mexico and he is said to have been the first mine superintendent of the Flor de Peña Mine, Lampazos de Naranjo, Nueva León, Mexico. According to him, this was a mine with complex zinc and lead minerals. From this mine, Legrand collected a piece of massive blende carrying a small quantity of a substance identified and first described as a new mineral species by J. Drugman & M. H. Hey (1932). During his professional career Legrand probably also resided in Douglas, Arizona (USA), a mining town on the Mexican border. It is unknown to the authors where and when Legrand passed away. However, when Drugman & Hey (*op.cit.*) published the first mineralogical description of legrandite, Legrand had already passed away some years before (end of the 1920s?). The material these authors investigated was kindly put at their disposal by Legrand's widow.

LEPERSONNITE-(Gd)

Jacques Lepersonne (1909-1997).

$CaGd_2(UO_2)_{24}(CO_3)_8Si_4O_{28}.60H_2O$

Orthorhombic.

Shinkolobwe, Haut-Katanga Province, Democratic Republic of the Congo.

Belgium: Royal Museum for Central Africa, Tervuren - Type: HT: RGM 13.781 (~50 mg).
USA: National Museum of Natural History, Washington - Type: T: 150.228

IMA N°: 81-036

Deliens, M. & Piret, P. (1982) Bijvoetite et lepersonnite, carbonates hydratés d'uranyle et de terres rares de Shinkolobwe, Zaïre. *The Canadian Mineralogist*, **20**, 231-238.

http://www.handbookofmineralogy.org/pdfs/lepersonnitegd.pdf

Spherule of radiating acicular yellow lepersonnite-(Gd) crystals with orange fourmarierite.
Shinkolobwe, Haut-Katanga Province, D.R. Congo.
Royal Museum for Central Africa (N° 14047).

Spherules composed of radiating acicular yellow lepersonnite-(Gd) crystals.
Shinkolobwe, Haut-Katanga Province, D.R. Congo.
Royal Belgian Institute of Natural Sciences (N° C 5763).

Spherule composed of radiating acicular yellow lepersonnite-(Gd) crystals. *Shinkolobwe, Haut-Katanga Province, D.R. Congo. Royal Belgian Institute of Natural Sciences (N° C 5763).*

Fan-like aggregates of small acicular lepersonnite-(Gd). Shinkolobwe, Haut-Katanga Province, D.R. Congo. L. Vandenberghe.

Dense spherules composed of acicular brownish yellow lepersonnite-(Gd) crystals. Shinkolobwe, Haut-Katanga Province, D.R. Congo. F. Coune (N° 2681).

Jacques Lepersonne was born in Mulhouse (France) on October 26, 1909. Following a classical education at the Atheneum in Liège, he enrolled at the University of Liège (ULg). In 1932 he graduated as a mining engineer and subsequently was appointed assistant to the ULg (1932-1933). Concurrently, he completed his studies of engineering geology (1934). In 1935, he entered the geology department of the Museum of Belgian Congo (know today as the Royal Museum for Central Africa, RMCA) in Tervuren. He served this institution for about forty years. When World War II started, Lepersonne was on duty for the RMCA in the former Belgian Congo. He decided to stay in Africa until the end of the hostilities. Meanwhile, he served as head of the Regional Geological Survey in Léopold-ville (present-day Kinshasa) and was concurrently attached to the War Mining Output Department (*Direction de la Production minière de Guerre*, DPMG) from 1942 through 1943. Upon his return to Belgium, in 1946, he combined the position of curator to the RMCA and that of head to the Mining Research Syndicate of the Lower and Middle Congo (BAMOCO) with a position as chief scientist in charge of missions to Kasai and Kwango. Having been in charge of the department of geology and mineralogy of the RMCA for quite some time, Lepersonne was finally appointed director of all geological sections of this institute in 1970. He retired from the RMCA in 1974. Lepersonne passed away in Etterbeek (Brussels) on August 10, 1997.

With the exception of a few papers on the geology of Belgium published during the first years of his professional career, Jacques Lepersonne's scientific contributions exclusively have reference to the regional geology of the present Democratic Republic of the Congo. Many of his publications were written in co-authorship with mining engineer Lucien Cahen. The latter joined the RMCA at Lepersonne's request, and was to become the director of this federal institution later on. The geological map of Zaire (a former name of the Democratic Republic of the Congo) at the scale 1:2,000,000 as well as the explanatory notice for this document, which published by Lepersonne the year he retired from active service, are still reference documents to all geologists working on or interested in the geological constitution and the history of this central African country.

Biography

Delhal, J. (1999) *Musée royal de l'Afrique centrale, Tervuren, Département de Géologie et Minéralogie, Rapport Annuel 1997-1998*, 43.

Alexandre, J. (2000) *Bulletin des Séances de l'Académie royale des Sciences d'Outre-Mer*, **46**, 341-344.

Delhal, J. & Delmer, A. (2000) *Annuaire de l'Académie royale de Belgique*, **CLXV**, 65-70.

Bibliography

Alexandre, J. (2000) *Bulletin des Séances de l'Académie royale des Sciences d'Outre-Mer*, **46**, 344-349 (selected papers).

Delhal, J. & Delmer, A. (2000) *Annuaire de l'Académie royale de Belgique*, **CLXV**, 70-79.

MARTHOZITE

Aimé Marthoz (1894-1962).

$Cu^{2+}(UO_2)_3(Se^{4+}O_3)_2O_2.8H_2O$

Orthorhombic.

Musonoi, Kolwezi, Lualaba Province, Democratic Republic of the Congo.

France:
Muséum national d'Histoire naturelle, Paris - Type T
Université de Paris - Type: HT 12252

IMA N°: 68-016

Cesbron, F., Oosterbosch, R. & Pierrot, R. (1969) Une nouvelle espèce minérale: la marthozite. Uranyl-sélénite de cuivre hydraté. *Bulletin de la Société française de Minéralogie et de Cristallographie*, **92**, 278-283.
Warin, R., www.agab.be/mineralogie/Katanga/katanga.html.

Marthozite is not mentioned in the Handbook of Mineralogy.

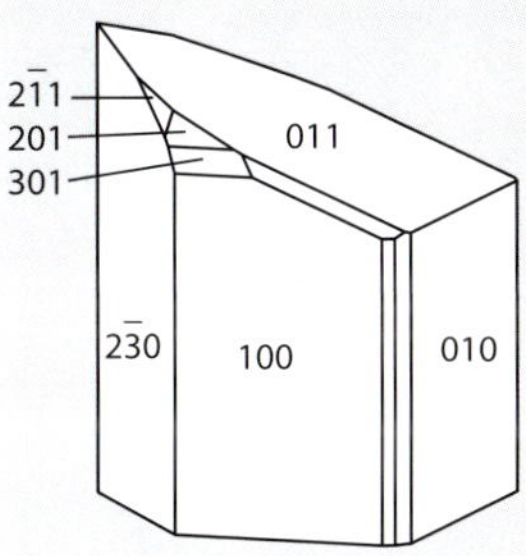

After Cesbron et al. (1969)

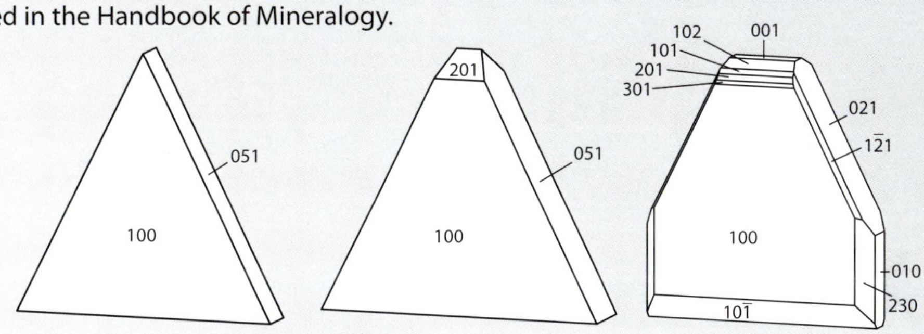

After Warin R. (November 2009)

Aggregate of khaki-coloured marthozite crystals. Musonoi, Lualaba Province, D.R. Congo. Royal Museum for Central Africa (N° 13231).

Parallelly grown translucent green marthozite crystals.
Musonoi, Lualaba Province, D.R. Congo.
P. du Ry (N° 1816).

Subparallelly grown marthozite crystals with typical triangular shape.
Musonoi, Lualaba Province, D.R. Congo.
J. Lhoest.

Aggregate of parallel grown, truncated triangular marthozite crystals.
Musonoi, Lualaba Province, D.R. Congo.
F. Coune (N° 3003).

Truncated triangular marthozite crystal.
Musonoi, Lualaba Province, D.R. Congo.
P. De Bondt.

Aimé Marthoz was born on September 13, 1894 in On (Marche-en-Famenne), in the Luxembourg Province. In 1912, he enrolled at Ghent University but his studies of construction engineering were strongly disturbed by the First World War. Because of the hostilities he had to interrupt his university education. He instead served as a technical agent in the Belgian Ardennes, where he was involved in projects concerning road construction and water distribution. At the close of the war, he resumed his studies in Ghent and graduated as a construction engineer in 1921. A few months after graduating, Marthoz was engaged by the *Union Minière du Haut-Katanga* (UMHK). After an appropriate training in Brussels, he left for Katanga where he was employed at the *Département des études et constructions.* As a result of his great merits and dynamic behaviour during the first years of his stay in Katanga, he was appointed head of this department by July 1931. Subsequently he became adjunct director general (1934) and director general for Africa (1937) of the UMHK in Elisabethville (now Lubumbashi).

During the beginning of the Second World War Marthoz was in Europe but returned to Katanga in 1942, where he worked at different production centres of the UMHK. During the post-war years he was appointed *administrateur-directeur* (1946) and managing director (1950). Around the same time he was asked to become a member of the Board of Directors of the *Société Générale de Belgique* (SGB). Thanks to his new senior functions, he managed to implement works devised and initiated years earlier, such as the construction of major quarries, factories, dams and power stations, and the automation of several factories. Concurrently, Marthoz got control over other societies working, or with interests in Katanga, such as the *Société métallurgique du Katanga*, the *Compagnie foncière du Katanga*, the *Société générale industrielle et chimique du Katanga*, the *Société générale des Forces hydro-électriques du Katanga*, the *Société générale africaine d'Electricité,* the *Compagnie maritime belge,* and the *Société générale métallurgique de Hoboken.* Two years after the Independence of the Congo, following a short illness, Marthoz died in Brussels on June 12, 1962.

Biography
Roger, E. (1963) *Bulletin des Séances de l'Académie royale des Sciences d'Outre-Mer*, **9**, 159-162.
Van der Straeten, E. (1973) *Biographie belge d'Outre-Mer*, **VII-A**, col. 337-340.
Kurgan-van Hentenrijk, G., Jaumain, S. & Montens, V. (eds) (1996) Dictionnaire des Patrons en Belgique. Les Hommes, les Entreprises, les Réseaux, 450-451. Bruxelles: De Boeck & Larcier, Dept. De Boeck Université.

Bibliography
Van der Straeten, E. (1973) *Biographie belge d'Outre-Mer*, **VII-A**, col. 340-341.

MASUYITE

Gustave Masuy (1905-1945).

$Pb(UO_2)_3O_3(OH)_2.3H_2O$

Monoclinic (pseudo-hexagonal).

Shinkolobwe, Haut-Katanga Province, Democratic Republic of the Congo.

Belgium: Royal Museum for Central Africa, Tervuren - Type: NT.

IMA: G

Vaes, J.F. (1947) Six nouveaux minéraux d'urane provenant de Shinkolobwe (Katanga) : la billietite (urane hydraté de Ba), la vandendriesscheite (uranate hydraté de Pb), la masuyite (uranate hydraté de Pb), la richetite (uranate hydraté (?) de Pb), la studtite (carbonate hydraté d'urane) et la diderichite (carbonate hydraté d'urane). *Annales de la Société géologique de Belgique,* **70**, B212-225.

Deliens, M. & Piret P. (1996) Les masuyites de Shinkolobwe (Shaba, Zaïre) constituent un groupe formé de deux variétés distinctes par leur composition chimique et leurs propriétés radiocristallographiques. *Bulletin de l'Institut royal des Sciences naturelles de Belgique, Sciences de la Terre,* **66**, 187-192.

Warin, R. (2009) http://www.agab.be/mineralogie/Katanga/masuyite.html.

http://www.handbookofmineralogy.org/pdfs/masuyite.pdf

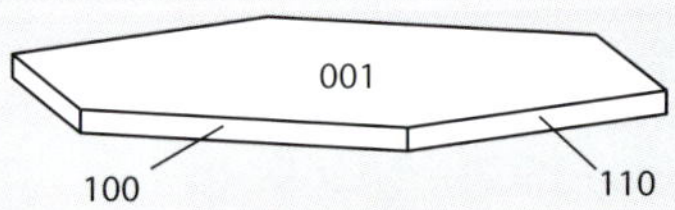

after Warin (2009)

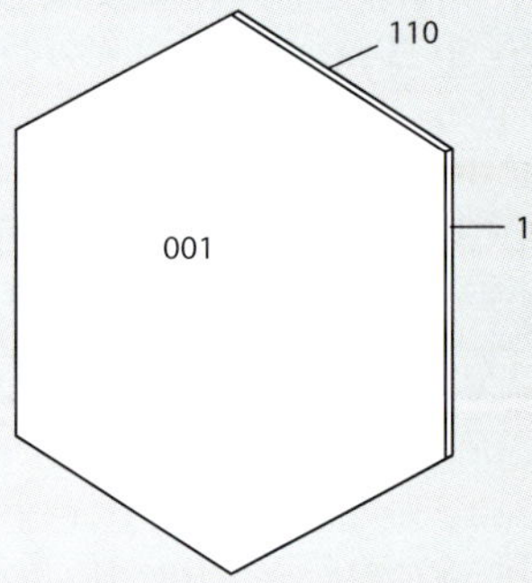

after Warin (2009) - rotated

Chemical analyses of more than 30 specimens belonging to numerous collections showed the presence of Pb in stoichiometric proportions. Pb was omitted in the original description (Michel Deliens, personal communication).

Spherical aggregates of acicular orange red masuyite crystals. Shinkolobwe, Haut-Katanga Province, D.R. Congo. R. van Dooren.

Tabular orange red masuyite crystals.
Shinkolobwe, Haut-Katanga Province, D.R. Congo.
Royal Museum for Central Africa (N° 13714).

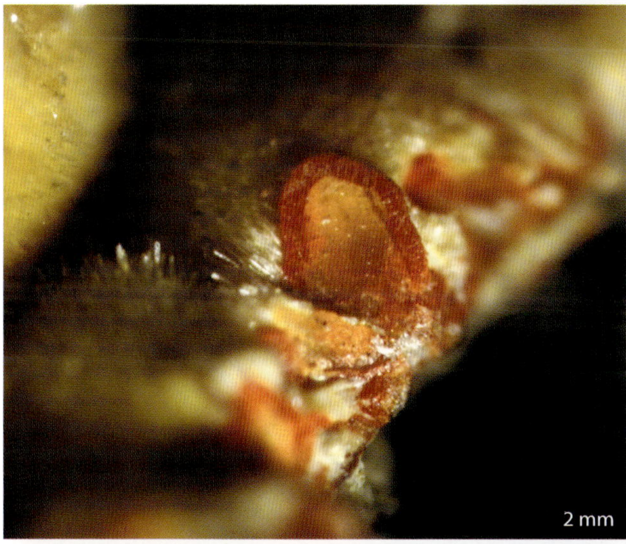

Tabular orange red masuyite crystal with thicker rim, and pale yellow acicular uranophane.
Shinkolobwe, Haut-Katanga Province, D.R. Congo.
L. Vandenberghe.

Tabular orange red masuyite crystal.
Shinkolobwe, Haut-Katanga Province, D.R. Congo.
P. De Bondt.

Double terminated tabular orange red masuyite crystal.
Shinkolobwe, Haut-Katanga Province, D.R. Congo.
P. De Bondt.

Gustave Masuy (Masui?) was born in Assebroek (Brugge) on February 9, 1905. He was a student at the Royal Cadets School and the Royal Military School in Brussels, and was promoted to sub-lieutenant of the Engineering Corps in 1923. As a young graduate, he served the Royal Military School as tutor for the courses on construction and geology. Subsequently, he enrolled at the Free University of Brussels. In 1935 he graduated as a mining engineer, and three years later he was awarded the title of engineering geologist (1938) with a dissertation on copper minerals from the Prince Leopold Mine (Kipushi) in Katanga. Following his university education and military service, Masuy chose to work for the colonial mining industry and joined the *Société Belge de Recherches Minières en Afrique* (REMINA).

At the beginning of World War II, being a member of the reserve officers' training corps, Masuy was called under arms. After the unfortunate end of the Campaign of 1940, he decided to play an active role in the intelligence service, an organization that became better known as the Secret Army. He was arrested by the Gestapo on December 1, 1942, and transferred to the camps of Breendonck, Papenburg, Grossrozen and Nordhausen. In 1945, he had the good fortune to be liberated by the American Army. He returned to Belgium. Yet, as a result of physical deterioration and affliction, he passed away in Brussels hardly nineteen days after his return (June 9, 1945). Other sources, however, claim that he was shot by the Germans.

Biography

Stevens, Ch. (1945) *Bulletin de la Société belge de Géologie, de Paléontologie et d'Hydrologie*, **54**, 141.

Bibliography

Stevens, Ch. (1945) *Bulletin de la Société belge de Géologie, de Paléontologie et d'Hydrologie*, **54**, 141-142.

MÉLONJOSEPHITE

Joseph Mélon (1898-1991).

$CaFe^{2+}Fe^{3+}(PO_4)_2(OH)$

Orthorhombic.

Angarf Sud, Plaine de Zenaga, Morocco.

Belgium: Laboratory of Mineralogy, University of Liège - Type: HT: 18.587.
France: Musée de Minéralogie, Ecole des Mines de Paris - Type: CT: (donated by André-Mathieu Fransolet, 1973).
USA: National Museum of Natural History, Washington - Type: CT: 128338 (donated by André-Mathieu Fransolet, 1973).

IMA N°: 73-012

Fransolet, A.-M. (1973) La mélonjosephite, $CaFe^{2+}Fe^{3+}(PO_4)_2(OH)$, une nouvelle espèce minérale. *Bulletin de la Société française de Minéralogie et de Cristallographie*, **96**, 135-142.

http://www.handbookofmineralogy.org/pdfs/melonjosephite.pdf

Fibrous mass of mélonjosephite.
Angarf-South pegmatite, Zenaga Plain, Morocco.
Laboratory of Mineralogy, University of Liège.

Small mélonjosephite crystals in phosphated rock.
Sandamab pegmatite, Usakos, Karibib District,
Erongo Region, Namibia.
F. Hubert (N° 11993).

SEM photographs of mélonjosephite crystals.
Angarf-South pegmatite, Zenaga Plain, Morocco.
Laboratory of Mineralogy, University of Liège.

Joseph Mélon originated from the West-Flandres Province, where he was born in Ieper on January 19, 1898. He received his secondary schooling in Tienen. In 1921, he obtained his PhD degree in mineral sciences at the University of Liège (ULg). From 1920 to 1922 he served the ULg as an *aide-préparateur* allocated to the department of mineralogy and crystallography. In 1922 Mélon became *chef de travaux*, a post he held until 1948. From 1948 until 1953 he was lecturer. In 1953 he was promoted to full professor. He kept this post until 1968, when he officially retired from the university. As emeritus professor Mélon remained in charge of the mineral collections of the ULg. He passed away in Liège on March 8, 1991, at the advanced age of 93.

Mélon laid the foundation for substantial educational reforms in crystallography and mineralogy. His scientific output is mainly focused on the following geological topics: the study of minerals from Belgium and the former Belgian Congo, the determinative method of Barker, the physical properties of bone and teeth, and problems related to the formation of caves and cave deposits. He was succeeded by André-Mathieu Fransolet (*see*: fransoletite and parafransoletite).

MOREAUITE

Jules Moreau (°1931).

$Al_3(UO_2)(PO_4)_3(OH)_2.13H_2O$

Monoclinic.

Kobokobo, Sud-Kivu Province, Democratic Republic of the Congo.

Belgium: Royal Museum for Central Africa, Tervuren - Type: HT: RGM 6601; CT 6197 and 6203.

IMA N°: 84-010

Deliens, M. & Piret, P. (1985) Les phosphates d'uranyle et d'aluminium de Kobokobo. VII. La moreauïte, $Al_3UO_2(PO_4)_3(OH)_2.13H_2O$, nouveau minéral. *Bulletin de Minéralogie*, **108**, 9-13.

http://www.handbookofmineralogy.org/pdfs/moreauite.pdf

SEM photograph of moreauite crystals. Kobokobo, Sud-Kivu Province, D.R. Congo. Royal Museum for Central Africa.

Aggregates of yellow moreauite crystals.
Kobokobo, Sud-Kivu Province, D.R. Congo.
Royal Museum for Central Africa. Holotype (N° 6201).

Aggregate of yellow moreauite.
Kobokobo, Sud-Kivu Province, D.R. Congo.
Royal Belgian Institute of Natural Sciences (N° 89511).

Globular aggregate of moreauite with furongite.
Kobokobo, Sud-Kivu Province, D.R. Congo.
Royal Museum for Central Africa (N° 6203).

Small yellow globules of moreauite crystals.
Kobokobo, Sud-Kivu Province, D.R. Congo.
Royal Museum for Central Africa (N° 6203).

Jules Moreau was born on January 28, 1931, in Nimy (Mons), in the Hainaut Province. Following his secondary studies at the Saint-Stanislas College in Mons, he set out to study at the *Faculté Polytechnique de Mons*. In 1954 he graduated as a mining engineer at the Catholic University of Leuven (CUL). As part of his university education, he travelled to the former Belgian Congo in 1953. In 1954-1955, funded by Fulbright and Esso grants, he sojourned in the USA and Canada. Moreau seized this opportunity to continue his education in the USA, obtaining the degree of Master of Arts, Geological Sciences, at Harvard University. Prior to his military service (1956-1957), he also spent six months at the geology department of the University of Amsterdam. His stay in The Netherlands was sponsored by a fellowship of the cultural relations between the Kingdom of Belgium and the Kingdom of The Netherlands.

When Jacques Thoreau (*see*: thoreaulite) retired from the CUL (1957), Moreau was promoted to lecturer. Four years later the Leuven alma mater offered him a full professorship. He lectured crystallography, mineralogy and metallogenesis in both linguistic sections of the university. In addition, he was in charge of the mineral and rock collections of the geology department. When the French-speaking section of the CUL moved to the campus of the *Université catholique de Louvain* in Louvain-la-Neuve (UCL), Moreau preferred to continue his education and research charges at the new location in the Walloon Brabant. Notwithstanding his full-time occupations at the UCL he accepted to continue assisting the students of the Dutch-speaking section in Leuven (*Katholieke Universiteit Leuven*). Here, he became an extraordinary professor in charge of the course on applied geology. Taking advantage of a sabbatical leave in 1974-1975, he spent one year at the Department of Earth Sciences of Purdue University, Lafayette, USA. Moreau retired from active service on October 1, 1996.

During his professional career, Moreau visited many mines in Europe, the USA, Canada, Morocco and the present-day Democratic Republic of the Congo. Among his research topics were studies of ore deposits, phase diagrams of sulphides of Cu, Fe and Ge, Nb-Ta oxide minerals, as well as kimberlites and diamond inclusions in rocks from the former Belgian Congo. For about thirty years, Moreau served as the Belgian representative to the Commission on New Minerals Nomenclature and Classification of the International Mineralogical Association (IMA-CNMNC).

Biography

Louvain-la-Neuve, Archives de l'UCL, dossiers du personnel.

MUNDITE

Walter Mund (1892-1956).

$Al(UO_2)_3(PO_4)_2(OH)_3.5H_2O$

Orthorhombic.

Kobokobo, Sud-Kivu Province, Democratic Republic of the Congo.

Belgium: Royal Museum for Central Africa, Tervuren - Type: HT: RGM 11.888; CT: 5 specimens.

IMA N°: 80-075

Deliens, M. & Piret, P. (1985) Les phosphates d'uranyle et d'aluminium de Kobokobo. V. La mundite, nouveau minéral. *Bulletin de Minéralogie*, **104**, 669-671.

http://www.handbookofmineralogy.org/pdfs/mundite.pdf

SEM photograph of platy mundite crystals.
Kobokobo, Sud-Kivu Province, D.R. Congo.
Royal Museum for Central Africa.

Crust of golden platy mundite crystals.
Kobokobo, Sud-Kivu Province, D.R. Congo.
Royal Museum for Central Africa. Holotype (N° 11888).

Walter Mund was born in Antwerp on January 22, 1892. His father originated from the Rhineland (Germany), while his mother was a niece of the famous German politician Ludwig Windthorst. At the completion of his secondary studies at the Holy Lady College of the Jesuits in Antwerp (1910), Mund entered the Catholic University of Leuven (CUL) and ontained his candidate degree in natural sciences in 1912. In 1914 he obtained a doctor degree in chemical sciences.

In August 1914, Mund left for Manchester (Great Britain), where he worked as a chemist at a food factory. In 1915, he became a war volunteer. Being of German parentage, he stayed at the internment camp of Vernon, the chief town of the Eure department (France) for a while. Yet he was put to work at the dye industry and educational matters. Subsequently, he served the 1st Grenadiers Regiment until the armistice. At the close of World War I, he was appointed assistant to the chair of industrial chemistry, headed by Gustave Delmarcel, in Leuven. Concurrently, he was employed shortly as a chemist by the *S.A. Industries chimiques de Wilsele* (Leuven), which tried to set up a dye factory. By the end of 1919, Mund was proclaimed a laureate of the Inter-University Contest and was awarded the Jean-Servais Stas Prize (*see: stasite*) of the Belgian Royal Academy.

In October 1920, following a visit to the laboratories of Marie and Pierre Curie in Paris, he was appointed lecturer (*chargé de cours*) to the Faculty of Sciences of the CUL, and full professor at the Institute of Agronomy, in charge of teaching physical chemistry and analytical chemistry. From 1923 onwards he also lectured physical chemistry at the *Ecoles Spéciales*. In 1925 Mund was released from the course on analytical chemistry at the Institute of Agronomy but instead accepted the succession of the chair of general physics at the Faculty of Sciences. During the 1930s and 1940s his teaching programme included, amongst others, classes of analytical mechanics and elements of theoretical physics. Mund, known as a pioneer with respect to medical applications of radioactivity, passed away in Leuven in the morning of August 15, 1956.

Biography

van Waeyenbergh, H. (1954-1956) *Annuaire de l'Université catholique de Louvain*, **91**-2, CXVI-CXXIII.
Bruylants, A. (1954-1956) *Annuaire de l'Université catholique de Louvain*, **91**-2, CXXIV-CXLVIII.
Huyskens, P. (1980) Florilège des Sciences en Belgique **II**, 493-505. Bruxelles: Académie royale de Belgique, Classe des Sciences.
Huyskens, P. (1982) *Annuaire de l'Académie royale de Belgique*, **CXLVIII**, 79-98.

Bibliography

Huyskens, P. (1982) *Annuaire de l'Académie royale de Belgique*, **CXLVIII**, 98-113.

OOSTERBOSCHITE

Robert Oosterbosch (1908-1992).

$(Pd,Cu)_7Se_5$

Orthorhombic.

Musonoi, Lualaba Province, Democratic Republic of the Congo.

France: Musée de Minéralogie, Ecole des Mines de Paris - Type: HT.

IMA N°: 70-016

Johan, Z., Picot, P., Pierrot, R. & Verbeek, T. (1970) L'oosterboschite $(Pd,Cu)_7Se_5$, une nouvelle espèce minérale, et la trogtalite cupro-palladifère de Musonoï (Katanga). *Bulletin de la Société française de Minéralogie et de Cristallographie,* **93**, 476-481.

http://www.handbookofmineralogy.org/pdfs/oosterboschite.pdf

Concentrate of oosterboschite.
Musonoi, Lualaba Province, D.R. Congo.
Royal Belgian Institute of Natural Sciences (N° C 4608); donated by J. Jedwab.

(See also microphotographs of verbeekite).

Robert Oosterbosch originated from of Etterbeek (Brussels), where he was born on April 1, 1908. He received his secondary education at the Jesuit College of Saint Michael in Brussels (1918-1924). During late 1925 he enrolled at the Catholic University of Leuven (CUL). His studies of mining engineering and engineering geology were completed in the early 1930s. Following graduation from the CUL in February 1932, he joined the *Union Minière du Haut-Katanga* (UMHK) and promptly started his geological activities and exploration work in Katanga (former Belgian Congo). At the onset of World War II he was drafted into the army, and assigned to the Belgian Expeditionary Forces (1940-1945). The end of the war marked his return to Africa, where he became director of the geology department of UMHK (1961). Three years later he promoted to head of the geological laboratory of this company. As such he was a top representative of UMHK at many meetings, conferences and symposia all over the world (Copenhague, Lusaka, Prague, Kinshasa, etc.) for many years. Oosterbosch served UMHK and subsequently, under President Mobutu Sese Seko, the state-owned mining company Gécamines (*Générale des Carrières et des Mines*) until he retired from active service by the end of 1974. He passed away on November 11, 1992, aged 84.

For many years Oosterbosch was involved with the development and exploitation of ore bodies in the Upper Katanga. Several investigations that were carried out by him in this region concerned the mineralogy and the metallogenesis of copper- and cobalt-bearing deposits. One example comprises the Kipushi Mine, located hardly a few hundred meters from the border with Zambia. It is a famous Zn-Pb-Cu deposit, known to be a major producer of zinc, cadmium and germanium in Africa. It was discovered by George Grey, the British leader of an exploration team sent out by Robert Williams of the Tanganyika Concessions Ltd at the close of the 19th century.

OSWALDPEETERSITE

Oswald Maurice Peeters (°1945)

$(UO_2)_2CO_3(OH)_2.4H_2O$

Monoclinic.

Jomac Mine, Brown's Rim, San Juan County, Utah, USA.

Belgium: Royal Belgian Institute of Natural Sciences, Brussels - Type: T: RC 5166

IMA N°: 2000-034

Vochten, R., Deliens, M. & Medenbach, O. (2001) Oswaldpeetersite, $(UO_2)_2CO_3(OH)_2$. $4H_2O$, a new basic uranyl carbonate mineral from the Jomac uranium mine, San Juan County, Utah, U.S.A. *The Canadian Mineralogist,* **39**, 1685-1689.

http://www.handbookofmineralogy.org/pdfs/oswaldpetersite.pdf

Aggregates composed of radiating yellow oswaldpeetersite crystals with bluish white gypsum.
Jomac Mine, Brown's Rim, San Juan County, Utah, USA.
Royal Belgian Institute of Natural Sciences. Holotype (N° C 5166).

Spherical aggregates of radiating micrometric prismatic oswaldpeetersite crystals.
Jomac Mine, Brown's Rim,
San Juan County, Utah, USA.
R. Vochten (N° 524514).

Oswald Maurice Peeters was born in Kessel-Lo (Leuven), in the Flemish Brabant Province, on March 2, 1945. He received his secondary education at the Royal Atheneum in Leuven. Following his elementary and secondary studies, he entered the Catholic University of Leuven (CUL). Here, he graduated with a licentiate degree in chemistry (1969) and obtained his PhD in sciences in 1977. During his PhD studies, Peeters was offered an assistantship to the laboratory for analytical chemistry and medicinal physicochemistry (1970) of the *Katholieke Universiteit Leuven* (KUL). Today he still continues to serve this laboratory. During his professional life he passed through different ranks of the scientific and academic careers. In 1998, he was finally promoted to senior lecturer. Nowadays, Peeters is a staff member of the laboratory for biocrystallography (formerly named laboratory for analytical chemistry and medicinal physicochemistry), at the Faculty of Pharmaceutical Sciences.

The scientific research of Peeters primarily focused on the following topics: (a) studies of ligands and metal complexes related to analytical techniques, (b) crystal structure determination, conformational analysis and structure-activity relationships of medicines, and, last but not least, (c) crystallographic studies in relation to mineralogy, with particular attention to unknown secondary uranium minerals. As far as his crystallographic research is concerned, Oswald Maurice Peeters co-authored two publications on new mineral species (piretite and deliensite) respectively discovered in Shinkolobwe, Katanga (Democratic Republic of the Congo), and at the Mas d'Alary uranium deposit, nearby Lodève, in the Hérault Department (France). These papers came about in close collaboration with Norbert Blaton (*see*: blatonite), Michel Deliens (*see:* deliensite) and Renaud Vochten (*see*: vochtenite).

PIRETITE

Paul Piret (1932-1999).

$Ca(UO_2)_3(Se^{4+}O_3)_2(OH)_4.4H_2O$

Orthorhombic.

Shinkolobwe, Haut-Katanga Province, Democratic Republic of the Congo.

Belgium: Royal Belgian Institute of Natural Sciences, Brussels - Type: HT: RC 4717.

IMA N°: 96-002

Vochten, R., Blaton, N., Peeters, O. & Deliens, M. (1996) Piretite, $Ca(UO_2)_3(SeO_3)_2(OH)_4.4H_2O$, a new calcium uranyl selenite from Shinkolobwe, Shaba, Zaïre. *The Canadian Mineralogist*, **34**, 1315-1322.

http://www.handbookofmineralogy.org/pdfs/piretite.pdf

SEM photographs of piretite crystals.
Shinkolobwe, Haut-Katanga Province, D.R. Congo.
Royal Belgian Institute of Natural Sciences.

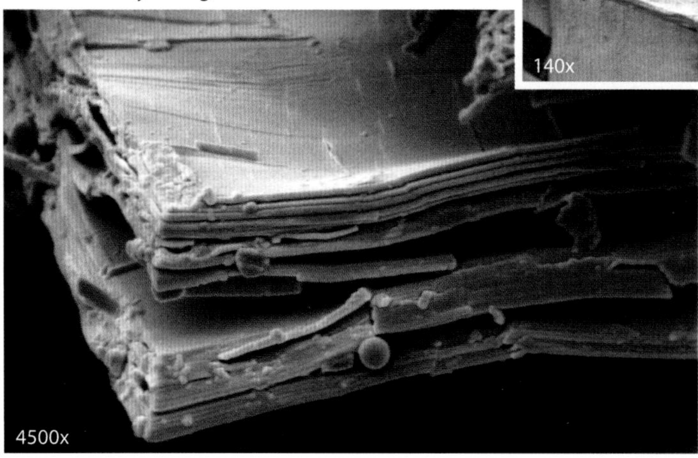

Aggregates of prismatic gold yellow piretite crystals.
Shinkolobwe, Haut-Katanga Province, D.R. Congo.
Royal Belgian Institute of Natural Sciences. Holotype (N° C 4717).

Fan-shaped aggregate of prismatic piretite crystals.
Shinkolobwe, Haut-Katanga Province, D.R. Congo.
R. Vochten (N° 472-7).

Hollow prismatic piretite crystals with rough surface.
Shinkolobwe, Haut-Katanga Province, D.R. Congo.
Royal Belgian Institute of Natural Sciences. Holotype (N° C 4717).

Hollow prismatic piretite crystals with rough surface.
Shinkolobwe, Haut-Katanga Province, D.R. Congo.
Royal Belgian Institute of Natural Sciences.

Paul Piret was born in Tournai, in the Hainaut Province, on May 13, 1932. In 1949 he completed his secondary schooling at the Saint Boniface College in Elsene (Brussels). The same year, with the prospect of performing engineering studies, he enrolled for a preparatory year at the Catholic University of Leuven (CUL). In 1956 he graduated with a doctor's degree in sciences (chemistry) at the CUL. About ten years later he became an *agrégé* Higher Education. Two years after earning his doctor's degree he attended a six-months-long practical training at the Belgian Nuclear Research Centre (SCK-CEN) in Mol. Subsequently he was awarded a post-doctoral research mandate by the Belgian National Fund for Scientific Research (NFWO-FNRS) (1958-1961). From 1961 to 1965 he served the same institution as a research associate. Between 1965 and 1971, he was adjunct university lecturer (*chargé de cours associé*) to the department headed by Maurice Van Meerssche (*see*: vanmeersscheite) at the *Université catholique de Louvain*, in Louvain-la-Neuve (UCL). Paul Piret promoted to associate professor (*professeur associé*) (1971-1974) and finally became full professor in 1974. As such, he continued to work at Louvain-la-Neuve until retirement from active service in 1997. He passed away in Ottignies on March 22, 1999.

During his professsional career, Paul Piret was much involved in crystallographic research, lectured crystallography, and contributed much to the systematic mineralogy of Congolese deposits. In close collaboration with Michel Deliens (*see*: deliensite) of the Royal Museum for Central Africa (Tervuren), and later serving the Royal Belgian Institute of Natural Sciences (Brussels), he studied e.g. the copper-cobalt- and uranium-bearing secondary mineral associations of southern Katanga (Shinkolobwe, Kamoto-East Cu-Co mine, etc.) and the uranium-bearing mineralizations associated with the beryl- and columbite-bearing granitic Kobokobo pegmatite (Lusungu River district, in Kivu, Democratic Republic of the Congo). In co-authorship with Deliens, Paul Piret described about thirty new mineral species (*see*: deliensite).

RENARDITE

Alphonse Renard (1842-1903).

$Pb(UO_2)_4(PO_4)_2(OH)_4.7H_2O$

Orthorhombic.

Shinkolobwe, Haut-Katanga Province, Democratic Republic of the Congo.

France: Muséum national d'Histoire naturelle, Paris - Type: T: 128.25.

IMA: Q

Schoep, A. (1928) La renardite, nouveau minéral uranifère. *Bulletin de la Société française de Minéralogie et de Cristallographie,* **51**, 247-252.

Deliens M., Piret, P. & Van Der Meersche , E. (1990) Les minéraux secondaires d'uranium du Zaïre. Deuxième complément, 39 p. Brussels: Institut royal des Sciences naturelles de Belgique; Tervuren: Musée royal d'Afrique Centrale.

Sejkora, J., Čejka, J. & Pauliš, P. (2003) Rare lead uranyl phosphate – dewindtite from the uranium occurrence Rýžoviště near Harrachov (Krkonoše Mts.), Bohemia, Czech Republic. *Bulletin mineralogicko-petrologického oddelení Národního muzea v Praze,* **11**, 177-183.

In 1928 Schoep described renardite as a new uranium mineral from Shinkolobwe. Deliens *et al.* (1990) stated that renardite was in fact a mixture of dewindtite and phosphuranilite. Sejkora *et al.* (2003) analysed specimens from Harrachov, Czech Republic and concluded that renardite is a valid mineral and not a mixture. Their conclusions did change the IMA status from discredited into questionable. More investigations are required to obtain a clear status.

Born in Ronse, in the East-Flanders Province, on September 28, 1842, **Alphonse Renard** received his secondary education in his birthplace and in Turnhout. In 1863 he entered the Jesuit noviciate in Drongen and took the vows in 1865. From 1866 to 1871 he was superintendent and taught at the Jesuit Colleges of Namur and Tournai. Meanwhile, he devoted himself to philosophy and natural sciences. Because of his interest and competence in these disciplines, he was sent to Maria Laach, in the Eifel (Germany). Here he familiarized himself with and deepened his knowledge on fieldwork. He also became interested in minerals and volcanic rocks. In 1872, the political situation in Germany forced Renard to return to Belgium, where he continued his geological and mineralogical research and set up a fruitful collaboration with Charles-Louis de la Vallée-Poussin of the Catholic University of Leuven. Meanwhile, he served as a professor of chemistry, geology and mineralogy in the Jesuit College in Leuven.

In 1877 Renard was ordained priest. During the summer of the same year he became curator of the department of mineralogy and lithology of the Royal Belgian Museum of Natural History (now Royal Belgian Institute of Natural Sciences, RBINS) in Brussels, a position he held until 1888. During his mandate at the mu-seum in Brussels and based upon the experience he gained in first-rate geological laboratories in Austria and Germany, Renard supplied his department with a laboratory for thin sectioning and wet chemical analysis of minerals and rocks, purchased optical instruments, and managed to engage Constantin Klement, a qualified chemist (*see*: *klementite*). The latter assisted him in quantitative chemical research and succeeded him when he left for Ghent University (GU). As time went by, Renard realized that it was impossible for him to reconcile the findings of modern scientific research and the dogmas of the Catholic Church. Therefore, in 1883 he turned his back upon his monastic order.

By the end of 1888, Renard joined GU where he became a full professor at the Faculty of Sciences. As such, he was responsible for training and research in crystallography, mineralogy, geology and stratigraphic palaeontology. A few years later, he also accepted to teach the same disciplines at the *Ecoles Spéciales du Génie Civil* in Ghent and started lecturing physical geography at the Faculty of Sciences. In 1901, he renounced both his faith and the laws of the Roman Catholic Church. After a long and painful illness, but pursuing his research to the very end of his life, Renard died on July 9, 1903, in Elsene (Brussels). With his death, GU lost one of its most prominent, influential and controversial professors.

Renard was the first Belgian to use the polarizing microscopy to study minerals and rocks. He acquired a very solid international reputation through his research on rock samples from oceanic islands and ocean floor sediments collected by the scientific staff of the *Challenger* expedition. During her scientific exploration voyage (December 1872 to May 1876), the Royal Navy corvette *HMS Challenger* criss-crossed all of the major oceans, with the exception of the Indian and the Arctic. Hence, his voyage forms the cornerstone of modern oceanography. The mineralogical and petrographical contribution of Renard to the *Challenger* research programme was published in a memoir dating back to 1888 entitled *'Report on the rock specimens collected on Oceanic Islands during the voyage of HMS Challenger'.* The results of his research on the dredged deep ocean sediments were embodied in the *'Report on the Scientific Results of the Voyage of HMS Challenger Deep Sea Deposits'* published in 1891 and written in collaboration with the Scottish-Canadian marine biologist Sir John Murray, the father of modern oceanography.

Renard and Murray also published in co-authorship the first map of the globe showing the regional distribution and the composition of the sediments capping the ocean floors. A first draft of the original map of 1894 is preserved and currently on display at the *Renard Centre of Marine Geology* (RCMG) in Ghent. This research centre is presently housed in the Department of Geology and Soil Science of GU, where it was informally installed in 1987.

Biography

Fiévez, Ch. (1904) Eloge fait à la Société belge d'astronomie (séance du 25 juillet 1903), 8 p. Bruxelles: Vve Ferdinand Larcier.
Université de Gand. Liber memorialis, Notices biographiques, **II**, 292-297, 1913. Gand: I. Vanderpoorten.
Buttgenbach, H. (1953) *Annuaire de l'Académie royale de Belgique*, **CXIX**, 387-404.
Stockmans, F. (1964) *Biographie nationale*, **32**, col. 604-608.
Roekeloos, A. (1976) *Annales du Cercle Historique et Archéologique de Renaix et du Tènement d'Inde*, **XXV**, 163-189.
Van Tassel, R. (1978) *Annales du Cercle Historique et Archéologique de Renaix et du Tènement d'Inde*, **XXVII**, 13-39.
Walschot, L. (1998) *Natuurwetenschappelijk Tijdschrift*, **78**, 288-292.
Blondeau, R.-A. (2005) *Nationaal biografisch woordenboek*, **17**, col. 528-533.

Bibliography

Buttgenbach, H. (1953) *Annuaire de l'Académie royale de Belgique*, **CXIX**, 405-413.
Université de Gand. Liber memorialis. Notices biographiques, **II**, 297-301, 1913, Gand: I. Vanderpoorten. w

RENIERITE

Armand Renier (1876-1951).

$(Cu^{1+},Zn)_{11}Fe_4(Ge^{4+},As^{5+})_2S_{16}$

Tetragonal (pseudo-cubic).

Prince Leopold Mine, Kipushi, Haut-Katanga Province, Democratic Republic of the Congo.

France: Musée de Minéralogie, Ecole des Mines de Paris.

IMA: G

Vaes, J.F. (1948) La reniérite (anciennement appelée "bornite orange"). Un sulfure germanifère provenant de la Mine Prince-Léopold, Kipushi (Congo belge). *Annales de la Société géologique de Belgique,* **72**, B19-32.

http://www.handbookofmineralogy.org/pdfs/renierite.pdf

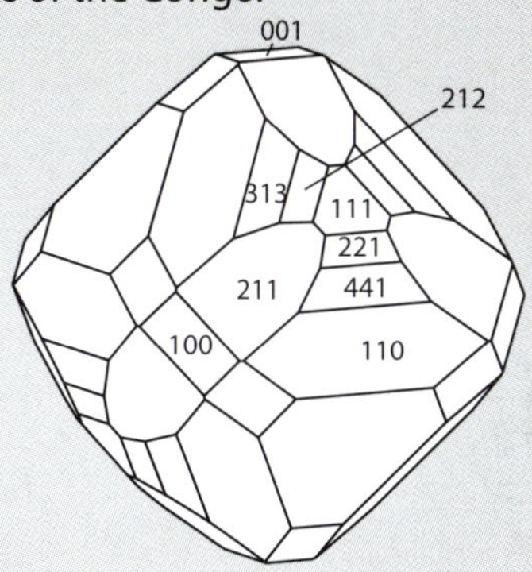

after Vaes (1948)

Renierite with typical orange bronze colour and metallic lustre.
Kipushi, Haut-Katanga Province, D.R. Congo.
Royal Belgian Institute of Natural Sciences.

RENIERITE

Renierite crystals on dolomite.
San Giovanni, Iglesias, Sardinia, Italy.
F. Hubert (N° 13860).

Armand Renier was born in Verviers, in the Liège Province, on June 26, 1876. He received his secondary education in his hometown and subsequently enrolled as a pre-engineering student at the University of Liège (ULg). In 1900 he graduated as a mining engineer and, after passing a proficiency examination, was recruited on probation by the *Corps des Mines*. Two years later he was appointed in this function. Concurrently he successfully finished his engineering geology studies. Based on the dissertation he submitted to this end he was declared a laureate of the *Concours des Bourses de voyage* of the Belgian Ministry for Education in 1902. This enabled him to visit Germany, the Austro-Hungarian Empire, Switzerland, the United Kingdom, France and Italy.

Early on in his professional career, Renier worked in various mining districts of Wallonia: Liège (1901-1903; 1906-1910), Mons (1905) and Charleroi (1905-1906). His administrative activities for the *Corps des Mines* were temporarily interrupted from November 1910 to May 1912. Because Julien Fraipont (*see*: fraipontite) died quite suddenly in the beginning of 1910, Renier accepted to teach his classes on 'Elements of vegetal palaeontology' at the Technical Faculty of the ULg. Renier's appointment was anything but surprising as since 1908 he had been tutor for courses on mine workings and palaeontology at the ULg. In mid-1912 he resumed his activities for the *Corps de Mines*. Yet, when shortly afterwards the first director of the Geological Survey of Belgium, i.e. Michel Mourlon, retired from active service, he succeeded him. Renier headed this institution seated in Brussels, until his official retirement (September 30, 1942). In 1931, Renier was appointed honorary professor to the ULg. Until 1946 he nevertheless held a limited teaching programme at this institution. Renier passed away in Brussels (?) on October 9, 1951.

The research activities of Renier comprised a very detailed study of the stratigraphy and the tectonics of the Coal Measures in Belgium. His observations and conclusions on this subject were brought together in a masterly work entitled *Les gisements houillers de la Belgique*. It is a synthesis of the main features and properties of these economically most valuable geological formations, and highlights their geographical distribution, palaeontological characteristics, stratigraphical structure, genesis, deformation history, useful minerals and ores, and some data on magnetism, seismology and hydrology. Another important part of Renier's research focused on palaeobotany. He did not limit his palaeobotanical studies strictly to scientific aspects, but paid attention to practical applications as well. Although Renier never did fieldwork in central Africa, he nonetheless wrote several papers on the coal formations of the former Belgian Congo.

Biography

Grosjean, A. (1951) *Bulletin de la Société belge de Géologie, de Paléontologie et d'Hydrologie*, **60**, 424-431.

Ancion, Ch. & Ubaghs, G. (1951) *Annales de la Société géologique de Belgique*, **75**, B7-9.

Fourmarier, P. (1956) *Annuaire de l'Académie royale de Belgique*, **CXXII**, 17-41.

Stockmans, F. (1964) *Biographie nationale*, **32,** col. 611-615.

Fourmarier, P. (1967) Florilège des Sciences en Belgique pendant le XIXe siècle et le début du XXe, 471-483. Bruxelles: Académie Royale de Belgique, Classes des Sciences.

Calembert, L. (1973) *Biographie belge d'Outre-Mer*, **VII-A**, col. 389-393.

Bibliography

Liber memorialis. L'Université de Liège de 1867 à 1935. Notices biographiques, **II**, 533-544, 1936. Liège: Rectorat de l'Université.

Grosjean, A. (1951) *Bulletin de la Société belge de Géologie, de Paléontologie et d'Hydrologie*, **60**, 431-436.

Fourmarier, P. (1956) *Annuaire de l'Académie royale de Belgique*, **CXXII**, 42-78.

Calembert, L. (1973) *Biographie belge d'Outre-Mer*, **VII-A**, col. 393-399.

RICHETITE

Emile Richet (1884-1938).

$(Fe^{3+},Mg)_x(Pb^{2+})_{8-6}(UO_2)_{36}O_{36}(OH)_{24}.41H_2O$

Triclinic.

Shinkolobwe, Haut-Katanga Province, Democratic Republic of the Congo.

Belgium: Laboratory of Mineralogy, University of Liège - Type: NT: 1712.

IMA: G

Vaes, J.F. (1947) Six nouveaux minéraux d'urane provenant de Shinkolobwe (Katanga): la billietite (urane hydraté de Ba), la vandendriesscheite (uranate hydraté de Pb), la masuyite (uranate hydraté de Pb), la richetite (uranate hydraté (?) de Pb), la studtite (carbonate hydraté d'urane) et la diderichite (carbonate hydraté d'urane). *Annales de la Société géologique de Belgique,* **70**, B212-225.
Piret, P. & Deliens, M. (1984) Nouvelles données sur la richetite $PbO.4UO_3.4H_2O$. *Bulletin de Minéralogie*, **107**, 581-585.

http://www.handbookofmineralogy.org/pdfs/richetite.pdf

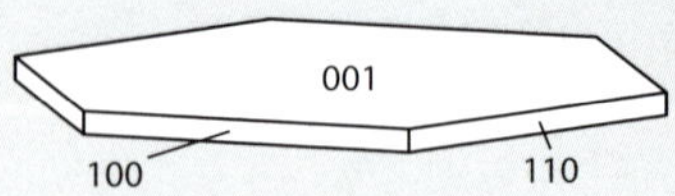

after Vaes (1947)

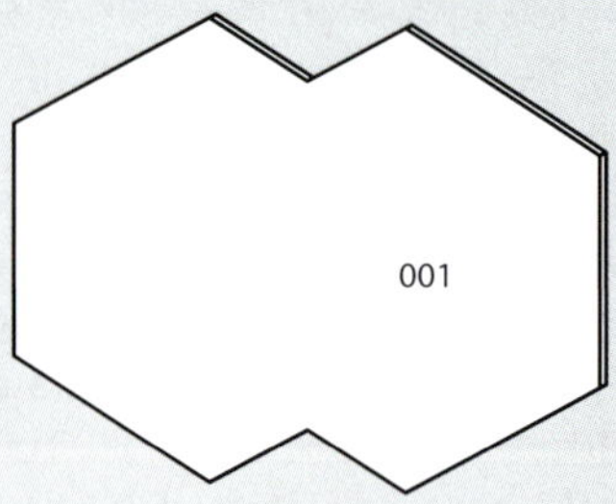

after Vaes (1947)

Reticulated aggregate of dark brown richetite associated with orange billietite crystals. Shinkolobwe, Haut-Katanga Province, D.R. Congo. Royal Museum for Central Africa (N° 13714).

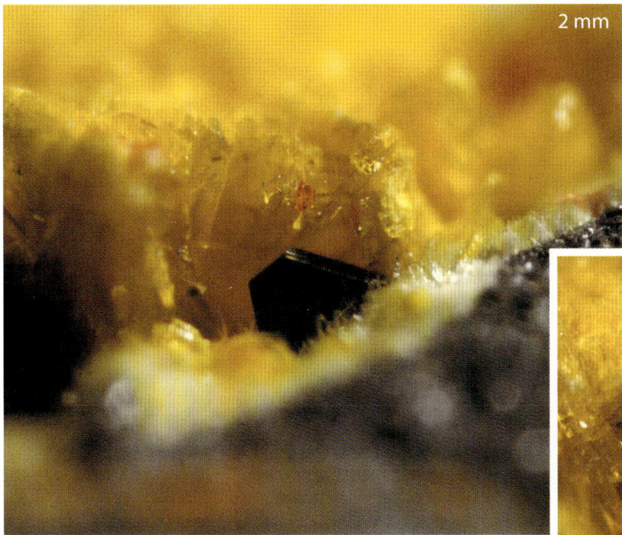

Shiny dark brown to black crystals of richetite on pale yellow uranophane and orange yellow becquerelite.
Shinkolobwe, Haut-Katanga Province, D.R. Congo.
R. van Dooren.

Emile Richet originated from the Hainaut Province, where he was born in Lessines on April 16, 1884. In 1908 he graduated as a mining engineer from the *Ecole des Mines de Mons*.

Richet spent he greater part of his professional career in Katanga, in the former Belgian Congo. According to the member lists of the *Annales de la Société géologique de Belgique*, Richet was employed in central Africa. Here, he worked for the *Société de Recherches minières du Bas-Katanga*, near Kikondja, for the *Syndicat d'études de gisements et métallurgie au Congo*, in Elisabethville (Katanga), and from 1920 until 1930 for the *Union Minière du Haut-Katanga* (UMHK), in Elisabethville and Panda (Katanga). At the beginning of his collaboration with UMHK he served as a geologist and later on as chief geologist of the geology department. His professional activities for UMHK in Katanga covered three periods: November, 1920 to May, 1922, February, 1923 to February, 1926, and finally November, 1926 to October, 1929.

After his second term with UMHK in Katanga, Richet took advantage of his stay in Belgium to pass examinations and to earn an engineering geology degree at the *Ecole des Mines de Mons*. Early on in 1930, Richet decided to return to Belgium for good. He continued to serve the UMHK at his office in Brussels and became adjunct to the head of the geological department in Belgium. From May 1, 1933 until late February 1934, Richet was involved in the training of future employees of the geological survey of UMHK in Katanga. He passed away in Brussels, on May 25, 1938, at the early age of 54.

Biography

Coosemans, M. (1955) *Biographie coloniale belge*, **IV**, col. 757.
Anonymous (1965): Jules Cornet, 1865-1929. Centenaire de sa naissance 1965, 141-142, 1965. Bruxelles: Robert Louis.

Bibliography

Anonymous (1965) Jules Cornet, 1865-1929. Centenaire de sa naissance 1965, 142. Bruxelles: Robert Louis.

SALÉEITE

Achille Salée (1883-1932).

$Mg(UO_2)_2(PO_4)_2.10H_2O$

Monoclinic (pseudo-tetragonal).

Shinkolobwe, Haut-Katanga Province, Democratic Republic of the Congo.

Belgium: *Université Catholique de Louvain*, Louvain-la-Neuve - Type: HT: K 1.812.

IMA: G

Thoreau, J. & Vaes, J.F. (1932) La saléeite, nouveau minéral uranifère. *Bulletin de la Société belge de Géologie, de Paléontologie et d'Hydrologie*, **42**, 96-99.

Piret, P. & Deliens, M. (1980) Nouvelles données sur la saléeite holotype de Shinkolobwe. *Bulletin de Minéralogie*, **103**, 630-632.

http://www.handbookofmineralogy.org/pdfs/saleeite.pdf

Aggregates of curved yellow green saléeite crystals. Shinkolobwe, Haut-Katanga Province, D.R. Congo. L. Vandenberghe.

Reticulated aggregate of lamellar yellow saléeite crystals. Shinkolobwe, Haut-Katanga Province, D.R. Congo. P. De Bondt.

Aggregats of curved pale yellow saléeite crystals. Shinkolobwe, Haut-Katanga Province, D.R. Congo. F. Coune (N° 2690).

Platy brown yellow saléeite crystals with green metatorbernite.
Shinkolobwe, Haut-Katanga Province, D.R. Congo.
Royal Belgian Institute of Natural Sciences (N° 3681).

Clear rectangular yellow saléeite crystals with green metatorbernite.
Shinkolobwe, Haut-Katanga Province, D.R. Congo.
J. Gryson.

Aggregate of elongate micaceous pale yellow saléeite crystals.
Prévinquières, Aveyron, France.
Royal Belgian Institute of Natural Sciences (N° C 3968).

Aggregate of tabular saléeite crystals.
Capoterra, Sardinia, Italy.
R. Vandenbosch.

Tabular saléeite crystal.
Le Temple, Deux Sèvres, France.
F. Hubert (N° 12348).

METASALÉEITE

$Mg(UO_2)_2(PO_4)_2.8H_2O$

Tetragonal.

Shinkolobwe, Haut-Katanga Province, Democratic Republic of the Congo.

IMA: G

Thoreau, J. & Vaes, J.F. (1932) La saléeite, nouveau minéral uranifère. *Bulletin de la Société belge de Géologie, de Paléontologie et d'Hydrologie*, **42**, 96-99.

Mrose, M.E. (1950) Studies of uranium minerals (III): Saléeite from Schneeberg, Saxony. *The American Mineralogist*, **35**, 525-530.

Piret, P. & Deliens, M. (1980) Nouvelles données sur la saléeite holotype de Shinkolobwe. *Bulletin de Minéralogie*, **103**, 630-632.

Suzuki, Y., Sato, T., Isobe, H., Kogure, T. & Murakami, T. (2005) Dehydration processes in the meta-autunite group minerals meta-autunite, metasaléeite, and metatorbernite. *American Mineralogist*, **90**, 1308-1314.

Mrose (1950) investigated saléeite specimens from Shinkolobwe (Thoreau & Vaes, 1932) and concluded that these actually consisted of metasaléeite (*lower-hydrate-saléeite*) and were tetragonal.
Piret & Deliens (1980) analysed the holotype specimen of Thoreau & Vaes (1932) and pointed out that this specimen, stored under normal hygroscopic and thermal conditions, is saléeite. They confirmed that as a result of variable conditions, some specimens dehydrate and probably turn into metasaléeite. Recent studies with respect to this problem were carried out by Suzuki *et al.* (2005).

Achille Salée was born in Spa, in the Liège Province, on September 6, 1883. Following his secondary schooling at his birthplace and at the Minor Seminary in Sint-Truiden, he studied at the Great Seminary in Liège. He was ordained priest in 1906. On the recommendation of Mgr. Jacques Laminne, the former director of the Minor Seminary in Sint-Truiden, Salée enrolled at the Catholic University of Leuven (CUL) where he obtained his doctor's degree in natural sciences in 1910. His doctoral thesis dealt with corals from Carboniferous limestones in Belgium and comprised a revision of the Genus *Caninia*. One year later, he was proclaimed laureate of the *Concours des Bourses de voyage* of the Belgian Ministry for Education. This scientific award enabled him to continue his palaeontological research in London and Paris.

Upon his return to Belgium (1913), Salée was appointed ordinary professor in Leuven. His teaching mandate included stratigraphical palaeontology, animal palaeontology, anthropology, and in a later stage, geology of Africa and colonial geography as well. By 1930, he became director of the geological institute of the Leuven alma mater. In 1921-1922 Salée had the opportunity to participate in a geological mission to Ruanda-Urundi (a former Belgian protectorate), where he assisted Fernand Delhaye (*see*: delhayelite). As part of this mapping survey, Salée also studied the region of the central African Great Lakes (Virunga district). After his first stay in central Africa, Salée spent much more time in the areas he visited with Delhaye, especially during the second half of the 1920s and in the early 1930s. During the first days of March 1932 Salée had a car crash near Astrida (modern Butare), in the south of Ruanda. Shortly after this accident (March 13, 1932), he passed away at the mission-house of Isavi, at the early age of 49.

As he authored many publications in the field of palaeontology, Achille Salée was predestinated to continue the paleaontological research of canon Henry de Dorlodot, who is considered to be the founder of this branch of geology at the CUL. However, he is best known to the geological community for the geological map (six sheets in total) of Ruanda-Urundi at the 1:200,000 scale, published in co-authorship with his colleague and close friend Delhaye.

Biography

Kaisin, F. (1932) *Revue des questions scientifiques* 4th series, **22**, 5-29.

Delhaye, F. (1933) *Bulletin des Séances de l'Institut Royal Colonial Belge*, **4,** 28-37.

Sluys, M. (1952) *Biographie coloniale belge*, **III**, col. 765-775.

Bibliography

Delhaye, F. (1933) *Bulletin des Séances de l'Institut Royal Colonial Belge*, **4**, 38-39.

Sluys, M. (1952) *Biographie coloniale belge*, **III**, col. 775-776.

SCHOEPITE

Alfred Schoep (1881-1966).

$(UO_2)_8O_2(OH)_{12}.12H_2O$

Orthorhombic.

Shinkolobwe, Haut-Katanga Province, Democratic Republic of the Congo.

Canada: Royal Ontario Museum, Toronto - Type: M 13072.

IMA: G

Walker, T.L. (1923) Schoepite, a new uranium mineral from Kasolo, Belgian Congo. *The American Mineralogist,* **8**, 67-69.

Buttgenbach, H. (1923) Nouvelles observations sur des cristaux de schoepite. *Annales de la Société géologique de Belgique,* **48**, 163-167.

Schoep, A. (1930) Les minéraux du gîte uranifère du Katanga. *Annales du Musée du Congo belge, Minéralogie, Géologie, Paléontologie, Série 1,* **1,** fascicule 1, 42 p.

Ungemach, H. (1929) Précisions cristallographiques sur quelques minéraux du Congo belge. *Annales de la Société géologique de Belgique*, **52**, C75-85.

http://www.handbookofmineralogy.org/pdfs/schoepite.pdf

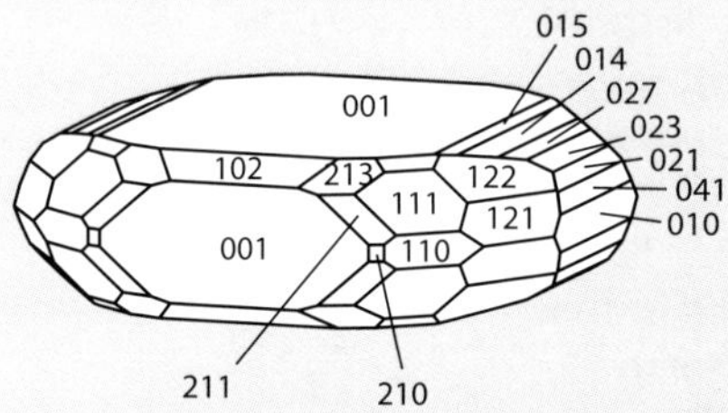

after Ungemach (1929)

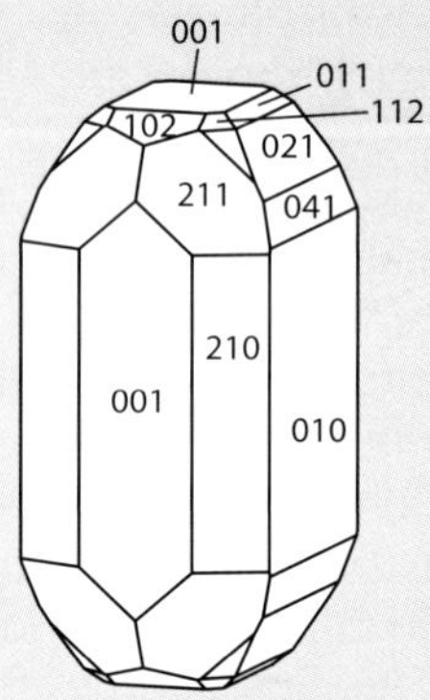

after Walker (1923)

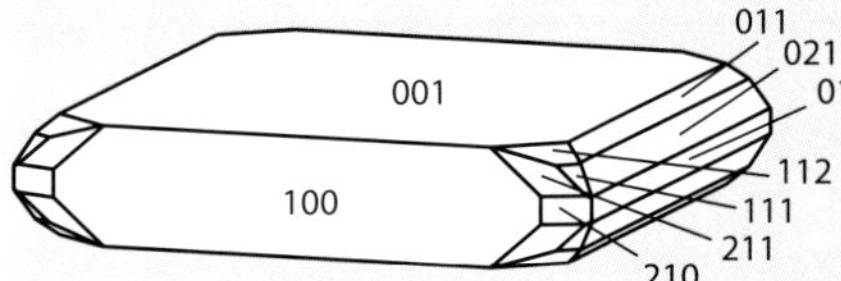

after Schoep (1930)

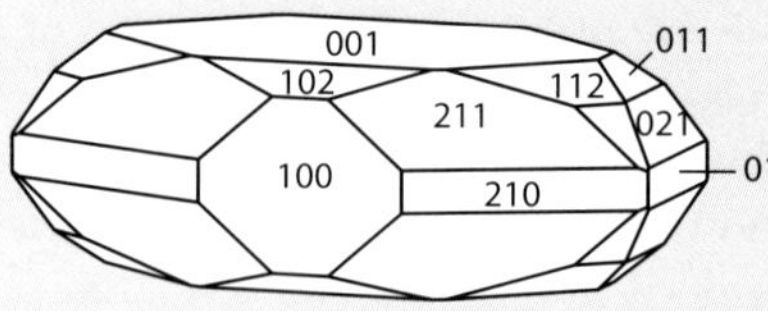

after Buttgenbach (1923)

Tabular brown yellow schoepite crystals.
Shinkolobwe, Haut-Katanga Province, D.R. Congo.
F. Coune (19237).

Clear tabular schoepite crystal.
Shinkolobwe, Haut-Katanga Province, D.R. Congo.
F. Hubert (N° 972).

Parallelly grown, tabular yellow schoepite crystals.
Shinkolobwe, Haut-Katanga Province, D.R. Congo.
L. Vandenberghe.

Tabular brown yellow schoepite crystals.
Shinkolobwe, Haut-Katanga Province, D.R. Congo.
F. Coune (19237).

PARASCHOEPITE

$UO_3.2\text{-}xH_2O$

Orthorhombic.

Shinkolobwe, Haut-Katanga Province, Democratic Republic of the Congo.

USA: National Museum of Natural History, Washington - Type: 94712.

IMA: G

Schoep, A. & Stradiot, S. (1947) Paraschoepite and epiianthinite, two new uranium minerals from Shinkolobwe (Belgian Congo). *The American Mineralogist*, **32**, 344-350.
Schoep, A. & Stradiot, S. (1948) Crystals of paraschoepite. *The American Mineralogist*, **33**, 513-514.

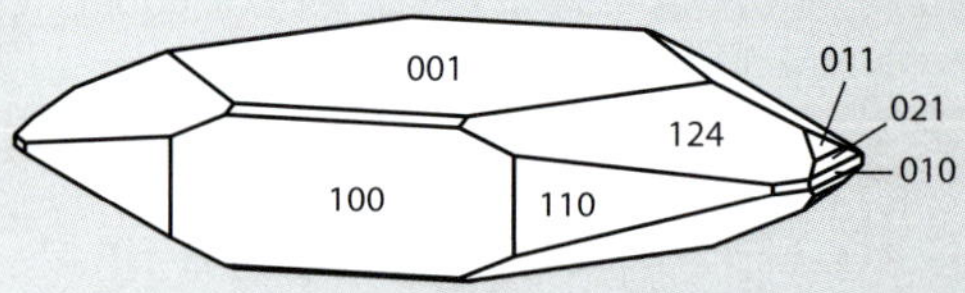

after Schoep & Stradiot (1947)

http://www.handbookofmineralogy.org/pdfs/paraschoepite.pdf

Prismatic paraschoepite crystals, with black coating (lower part of the picture).
Shinkolobwe, Haut-Katanga Province, D.R. Congo.
F. Coune.

Crust of yellow paraschoepite crystals.
Shinkolobwe, Haut-Katanga Province, D.R. Congo.
Royal Museum for Central Africa (N° 2707).

$(UO_2)_8O_2(OH)_{12}.10H_2O$

Orthorhombic.

Shinkolobwe, Haut-Katanga Province, Democratic Republic of the Congo.

USA: National Museum of Natural History, Washington - Type: 94712.

IMA: G

Christ, C.L. (1965) Phase transformations and crystal chemistry of schoepite. *The American Mineralogist*, **50**, 235-239.

Finch, R.J., Hawthorne, F.C. & Ewing, R.C. (1998) Structural relations among schoepite, metaschoepite and "dehydrated schoepite". *The Canadian Mineralogist*, **36**, 831-845.

Weller, M.T., Light, M.E. & Gelbrich, T. (2000) Structure of uranium(VI) oxide hydrate, $UO_3{\cdot}2H_2O$; synthetic meta-schoepite $(UO_2)_4O(OH)_6{\cdot}5H_2O$. *Acta Crystallographica*, **56**, 577-583.

http://www.handbookofmineralogy.org/pdfs/metaschoepite.pdf

Alfred Schoep was born in Ghent, in the East-Flanders Province, on June 6, 1881. Following his secondary education at his hometown, he attended the local university and completed his university studies obtaining the degrees of doctor in geography (1904) and doctor in mineral sciences (1906). His professional career began in 1906 when he was appointed assistant to the laboratory of analytical and toxicological chemistry of the Faculty of Medicine at Ghent University (GU). From 1910 to 1913, he participated in an expedition to Katanga (former Belgian Congo). This scientific mission was led by Constantin Guillemain, a geologist born and graduated at Breslau (present-day Wroclaw, Poland), who perished in September 1914 near Longwy, in Lorraine (France). When Guillemain passed away, Schoep performed field studies in the Caucasus, in particular in the Donetz region and in some areas near the Caspian Sea.

After World War I, Franz-Friedrich Stöber, who taught crystallography and mineralogy at Ghent University, resigned. Schoep succeeded him and became a member of the academic staff of both the Faculty of Sciences and the School of Technology. Concurrently, he started lecturing at the independent State College of Agriculture (now the Faculty of Bioscience Engineering of GU) in Ghent. From 1920 onward he was responsible for training and teaching general and applied geology at this college. In 1925 and during the following years, upon the advice and with the support of Jules Cornet (*see:* cornetite), Schoep was asked to teach mineralogy at the *Ecole des Mines de Mons* a task he accepted as well. In 1929, he was promoted to full professor and director of the mineralogy department of GU. As such, he became responsible for mineralogy, crystallography and the study of ore deposits. A few years later he also agreed to lecture applied geology and mining geography in the newly established section of mining engineering in Ghent.

The research and teaching duties of Schoep received a heavy blow when he was appointed administrator-inspector of GU (1932) and director of the Technical Schools. Thanks to his will-power, spirit of enterprise and ability for negotiation, he enabled many prestigious building projects of GU. This yielded his nickname 'Schoep-the-Builder'. The impact of Schoep's administrative tasks on his scientific activities was of such an extent that some tasks within the mineralogy department had to be taken over by Valère Billiet (*see*: billietite). At the end of the Second World War, Schoep was accused of having sympathized with the occupying forces and suspended from office. As a result, in 1946, the courses on mineralogy and petrography were assigned to Armand Hacquaert, crystallography was entrusted to Willy Dekeyser, whereas Georges Dewitte took on the task of teaching applied geology. Schoep was rehabilitated in the late 1940s and appointed emeritus professor in July 1951. After a long-lasting disease he passed away in Ghent on June 1, 1966.

Schoep gained broad international repute and recognition for his crystallographic studies on the secondary uranium minerals of Katanga. Over a period of about ten years (1921-1932), he discovered and authored the mineralogical characterization of sixteen new minerals, primarily originating from Shinkolobwe, Kasolo and Kalongwe. The present book sheds light on seven of these newcomers (buttgenbachite, dewindtite, dumontite, renardite, *stasite*, *stainierite*, vandenbrandeite). In the post-World War II era, the mineralogical community was informed about the discovery of two other new mineral species by Schoep: likasite (published in co-authorship with W. Borchert and K. Kohler) and paraschoepite (with Sadi Stradiot). Schoep was also one of the very first to provide geochronological data on the uranium minerals from southern Katanga. In February 1938, Schoep managed to show Herbert Clark Hoover - a former president of the USA and one of the first *doctores honoris causa* of GU - the department of mineralogy, when he paid a visit to GU. Taking into account that Hoover graduated in geology and mining engineering from Stanford University, USA, the discovery of a long series of new uranium minerals was without doubt a good reason for the organisation of this unique happening, a hallmark moment in the history of the Geological Institute of GU.

Biography

Hacquaert, A. (1940) 56^{e} Gentse Studentenalmanak - 't Zal Wel Gaan, 26-32.

Rijksuniversiteit te Gent. Liber memorialis 1913-1960, **IV**, 118, 1960. Gent: Uitgave van het Rectoraat.

De Leenheer, L. (1970) *The American Mineralogist*, **55**, 597-599.

Walschot, L. (1998) *Natuurwetenschappelijk Tijdschrift*, **78**, 293-294.

Bibliography

Rijksuniversiteit te Gent. Liber memorialis 1913-1960, **IV**, 119-122, 1960. Gent: Uitgave van het Rectoraat.

De Leenheer, L. (1970) *The American Mineralogist*, **55**, 599-602.

SENGIERITE

Edgar Sengier (1879-1963).

$Cu_2(UO_2)_2(VO_4)_2(OH)_2.6H_2O$

Monoclinic.

Luiswishi, Haut-Katanga Province, Democratic Republic of the Congo.

USA: Harvard University, Cambridge - Type: T: 103963.

IMA: G

Vaes, J.F. & Kerr, P.F. (1949) Sengierite: a preliminary description. *The American Mineralogist*, **34**, 109-120.

http://www.handbookofmineralogy.org/pdfs/sengierite.pdf

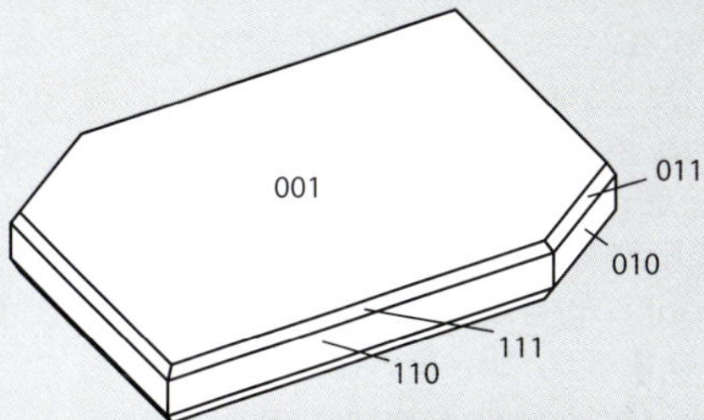

after Vaes & Kerr (1949)

Losange-shaped green sengierite crystal.
Luiswishi, Haut-Katanga Province, D.R. Congo.
J. Lhoest.

Rosette-like aggregates of platy sengierite crystals.
Musonoi, Haut-Katanga Province, D.R. Congo.
Royal Belgian Institute of Natural Sciences (N° C 3684).

Aggregate of green sengierite crystals.
Luiswishi, Haut-Katanga Province, D.R. Congo.
F. Coune (N° 1992).

Edgar Sengier was born in Kortrijk, in the West-Flanders Province, on October 9, 1879. He studied at the Catholic University of Leuven, and graduated as a mining engineer in 1903. The following year he practiced at a consulting company in Birmingham (U.K.), and concurrently improved his knowledge of English, accountancy and commercial correspondence. Subsequently he resumed his university studies and earned his electro-technical engineer degree in 1905. The same year he became manager of a coke-oven plant in Willebroek, in the Antwerp Province. Hardly two years later, under the authority of the *Compagnie internationale d'Orient,* he travelled to China. Here, he became adjunct administrator and subsequently administrator. As such, he took over from the Belgian engineer Jean Jadot who formerly built the famous Beijing-Hankow railway and later acquired fame in the former Belgian Congo (Jadotville, now Likasi, was named after him). During his four-year-stay in China, Sengier met businessmen from all over the world and gained great international experience. In 1911, he participated in a governmental commission of inquiry in Katanga. Back in Belgium (1912) he followed Jadot into the service of the *Union Minière du Haut-Katanga* (UMHK) in Europe.

When Belgium was invaded by the German forces in 1914, Sengier left for London. Here he had to lead the services of UMHK, which were at the time housed in the buildings of the Tanganyika Concessions Ltd. In London he met Herbert Hoover, president of the Commission for Relief in Belgium (CRB) and future president of the USA. By special request of Hoover, Sengier accepted to combine his job at the UMHK with a treasurership to the CRB. The close of World War I ended Sengier's activities in the United Kingdom. From then on he served the UMHK as director general for Africa (1918-1919). As such, he frequently visited Katanga during the interbellum and started a close collaboration with Jules Cousin (*see*: cousinite), who had great responsibilities in this region. From 1924, Sengier served as an admistrator and between 1932 and 1950 he was a managing director of the UM. Concurrently, he also occupied various positions in the *Société générale métallurgique de Hoboken* (now *Métallurgie Hoboken-Overpelt*) and the *Société Générale de Belgique* (SGB).

Shortly after World War II began, Sengier moved to New York where he represented UMHK, SGB, and their subsidiaries. Thanks to Sengier, who decided to transfer uranium ores from Katanga to New York in the early days of the war, the American Manhattan project, resulting in the production of the first atomic bombs, could be realized and the war was settled in favour of the Allies. No wonder thus that he was awarded the *Medal for Merit* by the United States government, the highest American distinction awarded to a non-American citizen for the first time. In post-war times, attaining an advanced age, Sengier gradually resigned from his administrative duties. At the close of 1960 and a few months after the Congo gained its independence, he retired as president of the Permanent Committee of UMHK. Sengier passed away in Cannes (France) on July 26, 1963.

Biography

Van der Straeten, E. (1973) *Biographie belge d'Outre-Mer*, **VII-A**, col. 429-437.

Goethals de la Kethulle de Ryhove, R. & Verheust, E. (1990) *Handelingen van de Koninklijke Geschied- en Oudheidkundige Kring van Kortrijk*, **LVI,** 195-233.

Kurgan-van Hentenrijk, G., Jaumain, S. & Montens, V. (eds) (1996) Dictionnaire des Patrons en Belgique. Les Hommes, les Entreprises, les Réseaux, 543-545. Bruxelles: De Boeck & Larcier, s.a. Dept. De Boeck Université.

STAINIERITE -> HETEROGENITE

Xavier Stainier (1865-1943).

Mindigi, Luisha and Kadjilangwe (Luambo), Haut-Katanga Province, Democratic Republic of the Congo.

IMA: D

Frenzel, A. (1872) Mineralogisches. 5. Heterogenit. *Journal für praktische Chemie*, **5**, 404-407.

Schoep, A. & Cuvelier, V. (1929) Sur la stainiérite (un hydroxyde cobaltique), nouveau minéral. *Bulletin de la Société belge de Géologie, de Paléontologie et d'Hydrologie*, **39**, 74-82.

Billiet, V. & Vandendriessche, A. (1939) Les oxydes hydratés de cobalt du Katanga. *Bulletin de la Société belge de Géologie, de Paléontologie et d'Hydrologie*, **49**, 63-78.

Hey, M.H. (1962) Cobaltic hydroxide in nature. *Mineralogical Magazine*, **33**, 253-259.

Hey (1962) gives a historical overview of the subsequent mineral names given by different authors since Frenzel (1872) attributed the name heterogenite to the cobaltic hydroxide he discovered at Schneeberg (Sachsen, Germany).
Ten years after the discovery and description of *stainierite* by Schoep & Cuvelier (1929), Billiet & Vandendriessche (1939) already proved that this mineral was not amorphous but in fact microcrystalline. They shed light on the position of *stainierite* but did not adopt the name heterogenite, described by Breithaupt more than sixty years earlier.
Among the more than twenty different names, used for cobaltic hydroxide, Hey (1962) mentions *transvaalite, stainierite, mindigite, trieuite* (*see: trieuite*) and *boodtite* (*see: boodtite*). Three of them are reminescent of Belgians.

Xavier Stainier was born on June 28, 1865, in Brye (Fleurus), in the Hainaut Province. After finishing his secondary education in Elsene and Brussels, he enrolled at the University of Liège (ULg) (1883). In 1888, he obtained his doctor's degree in natural sciences. His professional activities started when he was engaged as field geologist for the commission of the geological map of Belgium (1890). At the close of 1892, he was appointed lecturer to the *Institut agricole de l'Etat* in Gembloux, in charge of teaching geology and hydrology. The same year, Stainier became a member of the commission of the geological map of Belgium. When Alphonse Renard (*see*: ***renardite***) had passed away, Stainier was appointed an extraordinary professor of geology at Ghent University (GU). Five years later he became full professor, a position he filled until his retirement. His teaching programme included geology and physical geography for geology students, engineers and students of the Special School of Commerce. He combined his research and teaching duties at GU, with lecturing at the *Ecole des Hautes Etudes* in Ghent.

During World War I - when Dutch became the prevailing language at GU - Stainier was substituted temporarily (1917-1918) by Enno (Henk) Boeke, a Dutch geologist appointed full professor at the University of Frankfurt am Main in 1914. As he did not speak Dutch, Stainier was gradually relieved of his teaching duties in favour of Victor Van Straelen, when the changeover to Dutch at GU became a reality back in the 1920s. He became professor emeritus in 1935. On April 14, 1943 he died in Ghent.

Above all Stainier was a field geologist and an expert in mines. He spent countless days underground in mine-shafts. His favourite research subjects were stratigraphy and tectonics of the Belgian coal-basins. Though he is best known for his fundamental contributions on Coal Measures, he published with respect to nearly all research subjects in geological sciences. He also wrote papers on archaeological, prehistoric and anthropological topics. Stainier's bibliography mentions hundreds of papers and reports, representing over 6,000 pages of original scientific work. According to the bibliographic database established by the Geological Survey of Belgium, there are 312 bibliographic references to Stainier's name devoted to the geology of Belgium and published in leading Belgian geological journals. From this point of view, hitherto no other Belgian geologist did better than him. In addition, only a few papers in Stainier's complete bibliography were co-authored.

Biography

Université de Gand. Liber memorialis. Notices biographiques, **II**, 360-361, 1913. Gand: I. Vanderpoorten.
Stainier, X. (1939) Généologie de la famille Stainier. 254 p. Gand: L. Vanmelle.
Rijksuniversiteit te Gent. Liber memorialis 1913-1960, **IV**, 64-67, 1960. Gent: Uitgave van het Rectoraat.
Stockmans, F. (1970) *Biographie nationale*, **35**, col. 688-693.
Walschot, L. (1998) *Natuurwetenschappelijk Tijdschrift*, **78**, 302-303.

Bibliography

Université de Gand. Liber memorialis. Notices biographiques, **II**, 361-368, 1913. Gand: I. Vanderpoorten.
Rijksuniversiteit te Gent. Liber memorialis 1913-1960, **IV**, 68-79, 1960. Gent: Uitgave van het Rectoraat.

Jean-Servais Stas (1813-1891).

Kasolo , Shinkolobwe, Haut-Katanga Province, Democratic Republic of the Congo.

IMA: D

Schoep, A. (1922) Sur la stasite, un minéral nouveau, dimorphe de la dewindtite. *Comptes rendus de l'Académie des Sciences, Paris*, **174**, 875-877.

Schoep, A. (1923) Les minéraux uranifères (radioactifs) du Congo belge: becquerelite, curite, soddite, chinkolobwite, kasolite, dewindtite, parsonsite. *Bulletin de la Société belge de Géologie, de Paléontologie et d'Hydrologie*, **33**, 169-197.

In his description of *stasite,* Schoep mentioned that the mineral is orthorhombic, dimorphous with dewindtite, which he described as tetragonal.
Dewindtite was later shown to be orthorhombic, *stasite* and dewindtite thus being identical. As dewindtite was discovered before *stasite*, the latter name was discredited.

Jean-Servais Stas was born in Leuven, in the Flemish Brabant Province, on August 21, 1813. He studied at the university of his hometown and earned his title of doctor in medicine in 1835. When Laurent-Guillaume de Koninck (*see*: koninckite) left the Catholic University of Leuven (CUL), where he worked as a *préparateur* with Jean-Baptiste Van Mons (professor of chemistry, pharmaceutics and agronomy) Stas took his place. In 1835 Van Mons was appointed professor at Ghent University (GU). In the end, he did not take up this task due to his advanced age. In 1837 Stas decided to move to Paris, to work at the *Ecole Polytechnique* under the guidance of Jean-Baptiste Dumas. Back in Belgium, he accepted to lecture at the Royal Military School in Brussels in 1840. After the death of the physician Daniel Mareska, Stas was offered a professorship in Ghent. However, he preferred to concentrate on his full-time occupation in Brussels, where he was succeeded by Théodore Swarts in 1865. The professorship in Ghent was taken up by the German chemist August Kekulé von Stradonitz. When the latter left GU, Swarts became his successor in 1867. Stas agreed to re-occupy the post he previously held in Brussels. Hardly one year later Stas renounced all professional activities for health reasons. He passed away on December 13, 1891 in Saint-Gillis (Brussels).

The scientific work of Stas focused mainly on atomic weight determinations. This line of research rendered him world-famous. Under the guidance of and in collaboration with Dumas, he established e.g. the atomic weight of carbon. He also dedicated himself to the reformation of the university studies in Belgium and played a prominent part in the appointment of Kekulé as a full professor at the Ghent alma mater. Being a member of the Royal Academy of Belgium he also seated in the control commission for the geological map of Belgium. In 1850, Stas acquired a firm national reputation as chemist. Using an analytical method he designed himself, he successfully analysed the poison (nicotine) used by Count Hippolyte Visart de Bocarmé, known for his extravagant lifestyle, to kill his brother-in-law Gustave Fougnies. This sensational murder, that took place in the castle of Bitremont in Bury (near Mons) resulted in the public decapitation of the nobleman on the market-place of Mons in July 1851.

Biography

Spring, W. (1893) *Annuaire de l'Académie royale des Sciences, des Lettres et des Beaux-Arts de Belgique*, **59**, 217-360.

Spring, W. (1921-1924) *Biographie nationale*, **23**, col. 654-681.

Delhez, R. (1967) Florilège des Sciences en Belgique pendant le XIXe siècle et le début du XXe, 285-321. Bruxelles: Académie Royale de Belgique, Classe des Sciences.

Snelders, H.A.M. (2001) Jean-Servais Stas, een boegbeeld van de scheikunde. *In*: Halleux, R., Vandersmissen, J., Despy-Meyer, A. & Vanpaemel, G.: Geschiedenis van de wetenschappen in België 1815-2000, **1**, 155-158. Tournai: La Renaissance du Livre. Dexia Bank.

Bibliography

Spring, W. (1893) *Annuaire de l'Académie royale des Sciences, des Lettres et des Beaux-Arts de Belgique*, **59**, 361-372.

TAZIEFFITE

Haroun Tazieff (1914-1998).

$Pb_{20}Cd_2(As,Bi)_{22}S_{50}Cl_{10}$

Monoclinic.

Mutnovsky volcano, 70 km from Petropavlovsk-Kamchatskii, Kamchatka, Russian Federation.

IMA N°: 2008-12

Zelenski, M., Garavelli, A., Pinto, D., Vurro, F., Moëlo, Y., Bindi, L., Makovicky, E. & Bonaccorsi, E., (2009) Tazieffite, $Pb_{20}Cd_2(As,Bi)_{22}S_{50}Cl_{10}$, a new chloro-sulfosalt from Mutnovsky volcano, Kamchatka Peninsula, Russian Federation. *American Mineralogist*, **94**, 1312-1324.

Acicular black tazieffite crystals, reflecting greyish white, with colourful mutnovskite.
Mutnovsky volcano, Kamchatka, Russian Federation.
Institute of experimental mineralogy, Chernogolovska.
Photograph courtesy of Michael Zelenski.

Fissure with acicular black tazieffite crystals with a metallic lustre causing strong reflections.
Mutnovsky volcano, Kamchatka, Russian Federation.
G. Favreau.

Aggregate of radiating, tabular crystals of tazieffite.
Mutnovsky volcano, Kamchatka, Russian Federation.
Institute of experimental mineralogy, Chernogolovska.

Grass-shaped aggregates of tazieffite.
Mutnovsky volcano, Kamchatka, Russian Federation.
Institute of experimental mineralogy, Chernogolovska.

Tazieffite with platy cannizzarite on top of some aggregates
Mutnovsky volcano, Kamchatka, Russian Federation.
Institute of experimental mineralogy, Chernogolovska.

Tazieffite crystals with prismatic mutnovskite.
Mutnovsky volcano, Kamchatka, Russian Federation.
Institute of experimental mineralogy, Chernogolovska

All photographs courtesy of Michael Zelenski.

Haroun Tazieff was born in Warsaw (currently the capital of Poland) on May 11, 1914. Hardly a few months later his father, a Tartar Muslim doctor, was killed at the front. Together with his mother Haroun spent the first years of his life in Warsaw, Petrograd (Saint Petersburg) and Tiflis (Tbilisi), in the Caucasus. Subsequently they moved to Paris and later on to Belgium, where they settled down in Brussels in 1922. Haroun spent his earlier school days here and became a talented sportsman. Two years after he adopted the Belgian nationality, he graduated in agronomy at the *Faculté des Sciences Agronomiques de Gembloux* (1938). During World War II he served in the Belgian Army, joined the Resistance Movement and completed his studies of engineering geology at the University of Liège (1944). After graduation, he was offered assistantship at the Free University of Brussels (BFU). From 1945 to 1947 Tazieff worked for the *Service des mines du Katanga*. When his contract came to an end, he joined the *Service géologique du Congo belge*. During this period, he witnessed an adventive eruption of the Nyamuragira (the 1948 eruption of Mount Kituro). After experiencing the grandeur and fascination of this breath-taking event, the study of active volcanism became Tazieff's lifelong love and vocation.

After returning to Belgium, Tazieff took part in several expeditions. He accompanied e.g. the French undersea explorer Captain Jacques-Yves Cousteau when the *Calypso* headed for the Red Sea (1951-1952). He was also a member of the ill-fated exploration team that descended into the abyss of Pierre Saint-Martin in the Pyrenees (mid-1952). During the early 1950s Tazieff continued to serve as an assistant to the BFU. In the same period he travelled throughout the world to study active volcanoes and managed to set afoot the *Centre national belge de volcanologie* (1957) and the International Institute of Volcanological Research in Catania (Italy) (1962). From 1960 to 1975 he lectured volcanology at the Universities of Brussels, Paris Sorbonne, and Orsay.

In 1971 Tazieff was granted French citizenship. One year later he became research director to the *Centre National de Recherche Scientifique de France*. As such he headed the department of volcanology of both the *Institut de Physique du Globe* in Paris (1973-1976) and the *Centre des Faibles Radioactivités* in Gif-sur-Yvette (1977-1981). In 1981, Tazieff founded the Commision for the study and prevention of major natural hazards. Between 1984 and 1986, he was Secretary of State for the Prevention of Natural and Technological Hazards in the cabinet of the French Prime Minister Laurent Fabius. In his later years (1987-1996), Tazieff was president of the *Comité Supérieur d'Evaluation des Risques Volcaniques*, advising the French government. Death overtook him at home in Paris on February 2, 1998.

Thanks to his popular books, documentaries of rare beauty and compelling appearances on television, a wide audience was introduced to the world of volcanoes by Haroun Tazieff. He created a new kind of scientific communication, which was strongly criticized by some members of the academic community at the time.

Biography

McBirney, A. (1998) *Nature* 392 (6675), 244, April 2, 1998.

LeGuern, F. & Murray, J. (1998) *IAVCEI News* n° **1**, 6.

Tazieff, F. and coworkers (1998) Farewell, Tazieff. *CVS Newsletter* # **16**, 12-14, February 1998.

Cans, R. (1998) Tazieff, Le joueur de feu. Biographie. 286 p. Paris: Sang de la terre.

Tazieff, F. & Lacouture, J. (2004) Haroun Tazieff, Une vie de feu, 1-148. Grenoble: Glénat.

Bibliography

http://decobed.club.fr/tazieff.html

Tazieff, F. & Lacouture, J. (2004) Haroun Tazieff. Une vie de feu, 151. Grenoble: Glénat (list of books, films, documentaries and television series with the exception of the scientific publications).

THOREAULITE

Jacques Thoreau (1886-1973).

$Sn^{2+}Ta_2O_6$

Monoclinic.

Manono, Sud-Kivu Province, Democratic Republic of the Congo.

Belgium: Laboratory of Mineralogy, University of Liège - Type: T: 2067, 2068.

IMA: G

Buttgenbach, H. (1933) La thoreaulite, nouvelle espèce minérale. *Annales de la Société géologique de Belgique*, **56**, B327-328.

http://www.handbookofmineralogy.org/pdfs/thoreaulite.pdf

Yellow brown thoreaulite.
Manono, Sud-Kivu Province, D.R. Congo.
Royal Belgian Institute of Natural Sciences (N° C 4429).

Brown thoreaulite.
Manono, Sud-Kivu Province, D.R. Congo.
F. Coune.

Jacques Thoreau was born on September 27, 1886, in Elsene (Brussels). He received a classical education in Brussels. Afterwards, he enrolled at the *Ecoles Spéciales* of the Catholic University of Leuven (CUL) and graduated as a mining engineer in 1909. The following year, he was proclaimed laureate of the *Concours des Bourses de voyage* of the Belgian Ministry for Education. This award provided him with the opportunity to work at leading mining schools in Germany and to study metallogeny and mineralogy in Paris with Louis de Launay and Alfred Lacroix. At the same time he had the privilege of visiting mining districts in Germany and Austria. On his return to Belgium (1912), he served the CUL as a lecturer (*chargé de cours*) for the course on applied geology, and became a *suppléant* for the course on metallogeny.

In the early days of World War I, Thoreau became an extraordinary professor. When in late August 1914 his house went up in flames by acts of war, he left for Oxford. Here, he continued his scientific work and took advantage of his journey to prospect for oil in Spain (1916) and Mexico (1918) for the Belgian government. Back in Leuven, he became full professor (1920). He was in charge of teaching applied geology, crystallography, mineralogy and petrography. Concurrently with his academic tasks, Thoreau accepted to prospect in many non-European countries. This included the former Belgian Congo, where his main concern was hydrology, oil exploration and ore prospection. The aftermaths of his wife's early death (1930) and the onset of the Second World War brought his fieldwork nearly to a standstill, though by the end of the hostilities he resumed his field activities within and outside of Europe. From then on, parallel to his teaching and research at the university, he often worked for private colonial societies and public colonial organizations. His travels and scientific missions greatly enriched the collections of the geology department of the CUL. These collections were used to train students during his entire academic career. Thoreau retired and became a professor emeritus in 1956 but continued to spend his remaining time and efforts on scientific research. He passed away in Kessel-Lo (Leuven) on January 11, 1973.

Thoreau's scientific output primarily concerns the geology, mineralogy, petrography and metallogeny. It focuses on problems regarding central Africa and Katanga. By Thoreau's generous bequest, his mineral collection was left to the department of geology of the *Université catholique de Louvain* (UCL) in Louvain-la-Neuve. More recently, it was transferred to the *Musée d'Histoire naturelle* in Luxembourg, Grand Duchy of Luxembourg.

Biography

Moreau, J. (1961) *Mémoires de l'Institut géologique de l'Université de Louvain*, **XXII**, 1-2.

Denaeyer, M.-E. (1974) *Bulletin des Séances de l'Académie royale d'Outre-Mer*, **20**, 49-54.

Lepersonne, J. (1976) *Annuaire de l'Académie royale de Belgique*, **CXLII**, 51-58.

Thoreau, V. (1991) Jacques Thoreau. Un homme de science, un homme de foi, 119 p. Bruxelles: Didier Hatier.

Lederer, A. (1998) *Biographie belge d'Outre-Mer*, **VIII**, col. 418-420.

Delmer, A. (2007) *Nouvelle biographie nationale*, **9**, 343-344.

Bibliography

Denaeyer, M.-E. (1974) *Bulletin des Séances de l'Académie royale d'Outre-Mer*, **20**, 54-57.

Lepersonne, J. (1976) *Annuaire de l'Académie royale de Belgique*, **CXLII**, 58-66.

Robert du Trieu de Terdonck (1889-1970).

Star of the Congo Mine (l'Etoile du Congo Mine, Kulukulu Mine), Lubumbashi, Haut-Katanga Province, Democratic Republic of the Congo.

IMA: D

De Leenheer, L. (1935) Trieuiet, een nieuw Kobalt mineraal. *Natuurwetenschappelijk Tijdschrift,* **17**, 91-95.
Hey, M.H. (1962) Cobaltic hydroxide in nature. *Mineralogical Magazine,* **33**, 253-259.

For the history of *trieuite*, we refer to our remarks on *stasite*.

Robert du Trieu de Terdonck was born on September 25, 1889, in Muizen (Mechelen), in the Antwerp Province. He received his secondary education at the St. Rumbold's College in Mechelen and subsequently enrolled at the *Ecoles Spéciales* of the Catholic University of Leuven. From this institute he obtained degrees in mining engineering (1912) and engineering geology (1914). After graduating he shortly practiced at the construction workshops H. Bollinckx in Brussels, and at the *Charbonnages de l'Epine,* in Montignies-sur-Sambre (Charleroi). Subsequently, he was employed by the *Union Minière du Haut-Katanga* (UMHK), which he served until his retirement.

By mid-1914 du Trieu began working for the geology department of UMHK, in the former Belgian Congo. Here, because of the war, his first working period lasted until the end of April 1919. About one year later, he returned to Africa. In recognition of his outstanding services during the preceding years, he was promoted to head of the geology department. His first prolonged stay in Africa (1914-1919) was followed by two others, each three years long. In 1927, due to family affairs, du Trieu relinquished his post in central Africa, returned to Belgium and joined the central administration of the UMHK in Brussels. From now on, he adviced the department in the former Belgian colony he had previously directed. During four missions in the 1930s, he spent two more years in Africa. Thus, during his lifetime he resided a total of twelve years in the former Belgian Congo. During his professional career Robert du Trieu de Terdonck was above all active in the field of mine prospection. He retired from active service on December 31, 1959 and was succeeded by Hendrik Schuiling, a Dutch geologist. He passed away at his home in Brussels on December 9, 1970.

Biography

Staner, P. (1971) Mededelingen der Zittingen van de Koninklijke Academie voor Overzeese Wetenschappen/*Bulletin des Séances de l'Académie royale des Sciences d'Outre-Mer* **17**, 662-665.
Lederer (1998) *Biographie belge d'Outre-Mer*, **VIII**, col. 421-424.

Bibliography

Lederer (1998) *Biographie belge d'Outre-Mer*, **VIII**, col. 424.

VANDENBRANDEITE

Pierre Vanden Brande (1896-1957).

$CuUO_2(OH)_4$

Triclinic.

Kalongwe, Lualaba Province, Democratic Republic of the Congo.

France: Muséum national d'Histoire naturelle, Paris - Type: T: 134-72.
United Kingdom: National History Museum, London - Type: CT: 1933,261-263.

IMA: G

Thoreau J. (1931-1932) L'uranolépidite, nouveau minéral uranifère de Shinkolobwe, *Annales de la Société géologique de Belgique*, **55**, C3-5.
Schoep, A. (1932) La vandenbrandéite, un nouveau minéral uranifère. *Annales du Musée du Congo belge, Série I : Minéralogie*, **1**, 3, 23-31.
Milne, J.H. & Nuffield, E.W. (1951) Studies of radioactive compounds : I - vandenbrandeite. *The American Mineralogist*, **36**, 394-410.
Warin, R. (2009) http://www.agab.be/mineralogie/Katanga/vandenbrandeite.html.

http://www.handbookofmineralogy.org/pdfs/vandenbrandeite.pdf

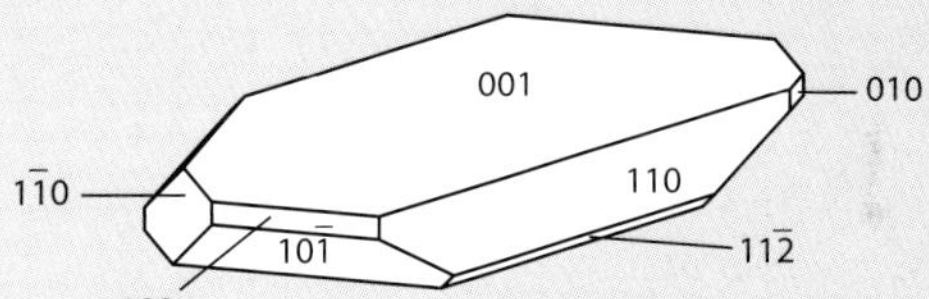

after Milne & Nuffield. (1951)

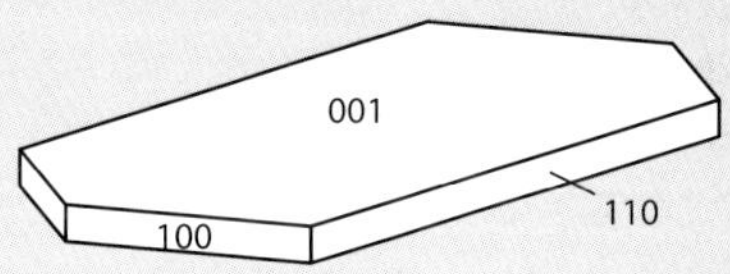

after Warin (2009)

One month before the publication of Schoep (1932), Thoreau (1931-1932) described the same mineral from Shinkolobwe as *uranolépidite*. As the description of Schoep (1932) was more complete, due to a better quality of the specimen, the new mineral received the name proposed by Schoep.

SEM photograph of vandenbrandeite crystals. Rabejac, Lodève, Hérault Department, France. Royal Belgian Institute of Natural Sciences.

Prismatic black green vandenbrandeite crystals.
Musonoi, Lualaba Province, D.R. Congo.
F. Coune (N° 3068).

Aggregate of radiating short-prismatic vandenbrandeite crystals.
Musonoi, Lualaba Province, D.R. Congo.
J. Lhoest.

Fan-like aggregate of tabular vandenbrandeite crystals.
Musonoi, Lualaba Province, D.R. Congo.
J.-M. Pendeville.

Tabular dark green vandenbrandeite crystals.
Musonoi, Lualaba Province, D.R. Congo.
J.-M. Pendeville.

Pierre Vanden Brande was born in Brussels on April 11, 1896. In 1919 he graduated in agronomical sciences at the *Institut agronomique de l'Etat* in Gembloux. In early 1920, he was employed by the *Comité Spécial du Katanga* (CSK). In Katanga (former Belgian Congo), he had a long and successfull professional career. Upon his arrival he was allocated to the *Service géographique et géologique* in Elisabethville (modern Lubumbashi), that was set up shortly before by the CSK and was headed at that time by Maurice Robert. During the late 1920s Vanden Brande also worked with the *Service des Mines* in Elisabethville for two years. Yet, afterwards he continued to serve the CSK. While engaged in Katanga his tasks were extremely diverse, including triangulation, topography, geological and pedological research (both in the field and in the laboratory), hydrology and applied soil science.

After the Second World War, Vanden Brande stayed in his homeland for about five years. In 1951 he decided to return to the former Belgian Congo to serve his former employer (CSK) again. In contrast to his first journeys in central Africa, Vanden Brande's research activities now focused upon soil survey and vegetation mapping in various districts of Katanga (Kokondja, Mitwaba, Kamina, Lubilash, etc.). On December 10, 1957, during soil and vegetation mapping activities nearby Kabombwe, he was suddenly attacked and mortally wounded by a herd of elephants.

Vanden Brande discovered the uranium mineral deposit of Kalongwe, in the Kolwezi area, Katanga. From this deposit, he collected the first sample of a new hydrous copper uranate that was first described and named after Vanden Brande by Alfred Schoep (*see*: schoepite) in 1932. Together with André Jamotte (1903-1951), a Belgian mining engineer and engineering geologist of the CSK killed during an aircraft crash in former British West Africa, Vanden Brande was also involved in the discovery of the Kengere site, in Katanga.

Biography

Grosemans, P. (1958) *Bulletin de la Société belge de Géologie, de Paléontologie et d'Hydrologie*, **67**, 296-298.
Walraet, M. (1968) *Biographie belge d'Outre-Mer*, **VI**, col. 112-113.

Bibliography

Grosemans, P. (1958) *Bulletin de la Société belge de Géologie, de Paléontologie et d'Hydrologie*, **67**, pp. 298-299.
Walraet, M. (1968) *Biographie belge d'Outre-Mer*, **VI**, col. 113-114.

VANDENDRIESSCHEITE

Adriaan Vandendriessche (1914-1940).

$Pb_{1.6}(UO_2)_{10}O_6(OH)_{11}.11H_2O$

Orthorhombic.

Shinkolobwe, Haut-Katanga Province, Democratic Republic of the Congo.

USA: Harvard University, Cambridge - Type: 106523.

IMA: G

Vaes, J.F. (1947) Six nouveaux minéraux d'urane provenant de Shinkolobwe (Katanga): la billietite (urane hydraté de Ba), la vandendriesscheite (uranate hydraté de Pb), la masuyite (uranate hydraté de Pb), la richetite (uranate hydraté (?) de Pb), la studtite (carbonate hydraté d'urane) et la diderichite (carbonate hydraté d'urane). *Annales de la Société géologique de Belgique*, **70**, B212-225.

http://www.handbookofmineralogy.org/pdfs/vandendriesscheite.pdf

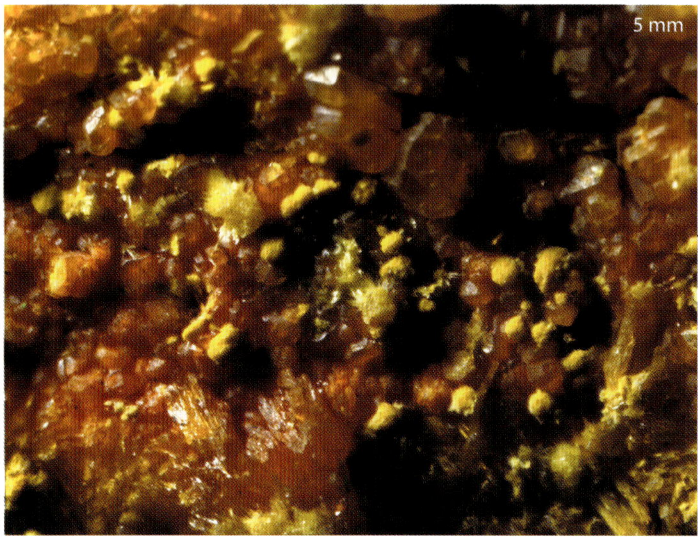

Dark orange vandendriesscheite crystals with yellow uranophane. Shinkolobwe, Haut-Katanga Province, D.R. Congo. L. Vandenberghe.

Prismatic dark orange vandendriesscheite crystals, yellow uranophane needles and grey rutherfordite nodules. Shinkolobwe, Haut-Katanga Province, D.R. Congo. Royal Museum for Central Africa (N° 2827).

Prismatic dark orange vandendriesscheite crystals. Shinkolobwe, Haut-Katanga Province, D.R. Congo. F. Coune (N° 3015).

METAVANDENDRIESSCHEITE

$PbU_7O_{22}.nH_2O$ (n < 12)

Orthorhombic.

Shinkolobwe, Haut-Katanga Province, Democratic Republic of the Congo.

USA: Harvard University, Cambridge - Type: 106523.

IMA: G

Christ, C.L. & Clark, J.R. (1960) Crystal chemical studies of some uranyl oxide hydrates [vandendriesscheite-II]. *The American Mineralogist*, **45**, 1026-1061.

http://www.handbookofmineralogy.org/pdfs/metavandendriesscheite.pdf

Adriaan Vandendriessche was born in Menen, in the West-Flanders Province, on January 13, 1914. Following his secondary studies in Kortrijk and Ghent, he enrolled at Ghent University (GU) (September 1931). He obtained his master degree in geology in July 1935. On November 21, 1938 he became laureate of the *Inter-University Contest* and obtained his doctor's degree in sciences, specializing in mineralogy and petrography, in October 1939. His doctoral work dealt with the petrography of metamorphic rocks from the Bastogne-Libramont area in the Belgian Ardennes, and was carried out under the inspiring guidance of Alfred Schoep (*see*: schoepite).

The professional career of Vandendriessche was very short. Following his military service, he was appointed superintendent and served as a teacher of natural sciences at the secondary boys' school of the State and the Atheneum in Kortrijk (October 1937). In his sparetime and for financial purposes, he lectured natural sciences at the *Rijksnormaalschool* (Teachers' Training College of the State) in Ghent, where his parents lived as well. Despite his demanding teaching job, Vandendriessche continued to do research at the laboratory of mineralogy of GU. His scientific work focused upon uranium minerals from Katanga, and metamorphic minerals and rocks from southern Belgium. Nearly half of his research publications were written in coauthorship with Valère Billiet (*see*: billietite).

By the end of August 1939, on the threshold of the Second World War, Vandendriessche was mobilized as an artillery-officer. During the 18 day campaign (May 1940), he was commander of the 4th Battery of the 7th Artillery Regiment and was killed in action at the hamlet called '*Konijn*' in Ursel (Knesselare). Struck in the breast by a salvo of a submachine gun, he died on May 27, 1940, at the age of 26. His mortal remains were transferred to the municipal cemetery of Ghent. To commemorate him, his doctoral dissertation was published in its entirety about one year after his death.

Biography

Schoep, A. (1940) *Natuurwetenschappelijk Tijdschrift*, **22**, 235-236.
Bleuzé, A. & Vanbossele, J. (1994) De Kortrijkse gesneuvelden 1940-1945, 42.

Bibliography

Schoep, A. (1940) *Natuurwetenschappelijk Tijdschrift*, **22**, 236.

VANMEERSSCHEITE

Maurice Van Meerssche (1923-1990).

$U(UO_2)_3(PO_4)_2(OH)_6.4H_2O$

Orthorhombic.

Kobokobo, Kamituga, Sud-Kivu Province, Democratic Republic of the Congo.

Belgium:
Université Catholique de Louvain, Louvain-la-Neuve - Type: HT: F 360 (a few mg).
Royal Museum for Central Africa, Tervuren - Type: CT: RMG13749.

IMA N°: 81-009

Piret, P. & Deliens, M. (1982) La vanmeersscheite $U(UO_2)_3(PO_4)_2(OH)_6.4H_2O$, et la métavanmeersscheite $U(UO_2)_3(PO_4)_2(OH)_6.2H_2O$, nouveaux minéraux. *Bulletin de Minéralogie*, **105**, 125-128.

http://www.handbookofmineralogy.org/pdfs/vanmeersscheite.pdf

SEM photograph of vanmeersscheite crystals.
Kobokobo, Kamituga, Sud-Kivu Province, D.R. Congo.
Royal Museum for Central Africa.

Distorted aggregates of foliated gold yellow vanmeersscheite.
Krunkelbach, Menzenschwand, Baden-Württemberg, Germany.
Royal Belgian Institute of Natural Sciences (N° C 4299).

METAVANMEERSSCHEITE

$U(UO_2)_3(PO_4)_2(OH)_6.2H_2O$

Orthorhombic.

Kobokobo, Kamituga, Sud-Kivu Province, Democratic Republic of the Congo.

Belgium: Royal Museum for Central Africa, Tervuren - Type: HT: RGM 13.749.

IMA N°: 81-010

Piret, P. & Deliens, M. (1982) La vanmeersscheite $U(UO_2)_3(PO_4)_2(OH)_6.4H_2O$, et la métavanmeersscheite $U(UO_2)_3(PO_4)_2(OH)_6.2H_2O$, nouveaux minéraux. *Bulletin de Minéralogie*, **105**, 125-128.

http://www.handbookofmineralogy.org/pdfs/metavanmeersscheite.pdf

ARSENOVANMEERSSCHEITE

$U(OH)_4[(UO_2)_3(AsO_4)_2(OH)_2].4H_2O$

Orthorhombic.

Menzenschwand, Southern Black Forest, Baden-Württemberg, Germany.

Germany: SMN-Stuttgart - Type : HT

IMA N°: 2006-018

Walenta, K & Theye, T. (2007) Arsenovanmeersscheit, ein neues Uranmineral von der Uranlagerstätte Menzenschwand im südlichen Schwarzwald. *Der Aufschluß*, **58**, 159-164.

Maurice Van Meerssche was born in Sint-Niklaas, in the East-Flanders Province, on April 18, 1923. He spent his childhood in Ath and La Louvière. In 1941, after completing his secondary schooling in La Louvière, he decided to study pharmaceutical sciences at the Catholic University of Leuven (CUL). Upon completion of his second year of universitary education, he switched to sciences and earned his licence degree in chemistry in 1946. Van Meerssche continued his university studies and obtained his PhD degree in chemistry in 1948. His doctoral thesis dealt with the quantitative study of the exchange hydrogen-deuterium under the influence of α-irradiation. It was carried out under Walter Mund's guidance (*see*: mundite).

Van Meerssche began his professional career at the *Institut National des Mines* at Pâturages in the Borinage, studying the combustion of methane. In 1949, he returned to Mund's laboratory in Leuven, where he performed research under the sponsorship of the Belgian National Fund for Scientific Research (NFWO-FNRS). Concurrently, he accepted to teach physical chemistry at the *Institut Meurice Chimie* in Brussels and broadened his knowledge of quantum mechanics and molecular physical chemistry. During the early 1950s, Van Meerssche left for Utrecht, wto study structural radiocrystallography under chemistry Nobel Prize winner Johannes Martin Bijvoet. Thoroughly acquainted with this new line of research, he returned to the Leuven alma mater in 1954 and set up a laboratory for structural crystallography within the department of physical chemistry. He also agreed to lecture crystallography, as Jacques Thoreau (*see*: thoreaulite), who was in charge of this field of study, retired shortly before. Van Meerssche was also asked to assist Mund in his educational duties.

As a consequence of the death of Mund, Van Meerssche became responsible for the course on physical chemistry in 1958 and was appointed director of the laboratory of physical chemistry. During the following years, his research laboratory became increasingly involved in studies relating to molecular physical chemistry. Van Meerssche paid close attention to structural analysis and over the years extended his field of research to the nuclear magnetic resonance technique. When the *Université catholique de Louvain* (UCL) was founded, he decided to leave Leuven in favour of the French-speaking campus in Louvain-la-Neuve. One of his colleagues here was Paul Piret (*see*: piretite). Van Meerssche also lectured crystallography for about ten years at the Free University of Brussels. In 1989 he was appointed professor emeritus. On March 16, 1990, hardly a few months later, he passed away at Ottignies-Louvain-la-Neuve.

Biography

Reisse, J. (1995) *Annuaire de l'Académie royale de Belgique*, **CLXI**, 61-66.
Declercq, J.-P. (2005) *Nouvelle biographie nationale*, **8**, 374-375.

Bibliography

Reisse, J. (1995) *Annuaire de l'Académie royale de Belgique*, **CLXI**, 66-100.

VARLAMOFFITE

Nicolas Varlamoff (1910-1976).

$(Sn,Fe)(O,OH)_2$

Tetragonal.

Kalima, Maniema Province, Democratic Republic of the Congo.

IMA: Q

Buttgenbach, H. (1947) Les minéraux de Belgique et du Congo belge, 573 p., Paris: Dunod.
Sidorenko, G.A., Korovushkin, V.V., Gorshkov, A.I., Rudnitskaya, E.S., Dikov, Yu.P., Kaukova, L.V. & Sivtsov, A.V. (1993) The mineral nature of varlamoffite (in Russian). *Mineralogicheskii Zhurnal, Kiev*, **15**(4), 94-101.
Jambor, J.L., Tertsev, N.N. & Roberts, A.C. (1995) New Minerals, *American Mineralogist*, **80**, 845-850.

Buttgenbach (1947) presented unpublished data received from R. De Dycker.
Jambor *et al.* (1995) concluded that the results of Sidorenko *et al.* (1993) seemed to validate a previously ill-defined species and should be submitted to the CNMMN of IMA for a formal redefinition of the mineral.

Ochre yellow varlamoffite in quartz.
Pangi, Munkuku, Messaraba, Sud-Kivu Province, D.R. Congo.
Royal Museum for Central Africa (N° 7976).

Ochre yellow varlamoffite in quartz.
Pangi, Munkuku, Messaraba, Sud-Kivu Province, D.R. Congo.
Royal Museum for Central Africa (N° 3227).

Nicolas Varlamoff was a Don Cossack born in Stanitza Krasnokoudskaya (southern Russia) on April 12, 1910. In 1920, the Varlamoff family sailed from the Crimea to Turkey, and took up residence in Constantinopel (nowadays Istanbul) til 1922. Subsequently, the Varlamoff's moved to Belgium. The young Nicolas received his secondary education in Malonne, in the Namur Province. Following his secondary schooling he enrolled at the University of Liège (ULg) and graduated as a mining engineer in 1934. Two years later he obtained his engineering geology degree at the same institution. After completing his university studies, he was restrained from leaving for Africa by a severe illness. Instead, he accepted a director position at the *Carrières Unies de Grès Quevy Frères*, in Gives/Ben-Ahin, in the Liège Province (1936-1938).

In 1938 Varlamoff was employed by the *Société Belge de Recherches Minières en Afrique* (REMINA). This marked the outset of his colonial career. As time went by he served REMINA as *chef de secteur*, *chef de mission*, *ingénieur principal* and director for Africa. At the same time, he also worked as a consultant geologist with the *Syndicat Minier d'Etain* (SYMETAIN). Apart from pegmatite research, his professional activities in the former Belgian Congo mainly dealt with tin and tungsten exploration in the Maniema region (Kivu). He also studied other areas, mainly situated in Ruanda, Urundi and Madagascar. From 1962 onwards Varlamoff served as a technical advisor and representative of the United Nations Centre for Natural Resources, Energy and Transport in New York, N.Y., USA. This provided him with more missions in and outside Africa. In spite of his very demanding professional duties, he concurrently taught economic geology at the University of Brasilia, Brasil. Here he unexpectedly died on April 10, 1976.

Nicolas Varlamoff was a Cossack to the core, born a Russian citizen. Between his departure from Russia and the early 1950s he was a stateless person (Nansen Passport). He received the Belgian nationality in 1951. His death was announced in the first volume of the journal *Catastrophist Geology* under the telltale heading of '*Big Jaw Varlamoff is Dead*'. Indeed, his imposing and robust physical appearance and his tendency to recount tall stories, earned him the nickname of *Mashavu* (the great jaw) during his employment in the former Belgian Congo. For example, he took pride in telling everybody that as a student at the ULg, he worked as a roofer at the former Guillemins railway station in Liège to earn his living.

Biography

Tillé, R. (1977) *Bulletin des Séances de l'Académie royale des Sciences d'Outre-Mer*, **23**, 49-50.

Bibliography

Tillé, R. (1977) *Bulletin des Séances de l'Académie royale des Sciences d'Outre-Mer*, **23**, 50-52.

VERBEEKITE

Theodoor Verbeek (1927-1991).

$PdSe_2$

Monoclinic.

Musonoi, Lualaba Province, Democratic Republic of the Congo.

United Kingdom: National History Museum, London - Type: HT BM2001, 4.

IMA N°: 2001-005

Roberts, A.C., Paar, W.H., Cooper, M.A., Topa, D., Criddle, A.J. & Jedwab, J. (2002) Verbeekite, monoclinic $PdSe_2$, a new mineral from the Musonoï Cu-Co-Mn-U mine, near Kolwezi, Shaba Province, Democratic Republic of Congo. *Mineralogical Magazine*, **66**, 173-179.

http://www.handbookofmineralogy.org/pdfs/verbeekite.pdf

Photomicrographs of verbeekite in plane-polarized light (left) and between crossed polars (right).
On the image to the right, verbeekite is bronzy brown and the associated oosterboschite shows lamellar twinning and is grey.
Musonoi, Lualaba Province, D.R. Congo.
Mineralogical Magazine, ***66****, 173-179. Photograph courtesy of Jacques Jedwab.*

Theodoor Verbeek was born on September 1927, in Ophoven (Kinrooi), in the Belgian Limburg Province. He attended the local elementary school (1933-1940) and subsequently received a classical education in the nearby city of Maaseik (1940-1946). For his higher education he enrolled at Ghent University (GU). From 1946 to 1951 Verbeek studied pharmaceutical sciences. After graduation, he enrolled at the Faculty of Sciences to obtain a 'licence' diploma in geological sciences (1951-1954). His thesis was carried out under Victor Van Straelen and dealt with rock samples from the southern North Sea basin. Upon completion of this study, he decided to specialize in petrography at the laboratory of Maurice Roques, University of Clermont-Ferrand, in central France (1955).

Between 1955 and 1958 Verbeek performed extensive field research in the Congo Basin (in particular at Dekese, Kasai, Kwango, Aruwimi and Lindi) under the authority of the *Société Belge de Recherches Minières en Afrique* (REMINA). Subsequently he did laboratory research at both the Royal Museum for Central Africa (Tervuren) and the laboratory of general geology and petrography in Ghent. Verbeek submitted his doctoral thesis at GU. With this thesis under the guidance of Armand Hacquaert he obtained his PhD degree in 1960. His dissertation focused on the geology and the lithology of the Lindi Group, in the northeastern part of the former Belgian Congo.

At the beginning of his professional career (1960-1967), Verbeek worked as a geologist-mineralogist for the *Union Minière du Haut-Katanga* (UMHK) and the *Société Générale des Minérais* (SGM). As such he worked in the Tenke-Fungurume area, in the copper belt of Katanga, Congo. From 1967 until his retirement in 1983, he was an exploration manager for the *Union Minière Explorations and Mining Corporation Ltd* (UMEX) in Canada. As part of his field and laboratory research, he participated in many prospection campaigns in Quebec and in the northern part of Ontario (Thierry Mine, etc.). He also initiated several prospection programs in Mexico, Brazil and the USA. From 1983 until his demise, Theodoor Verbeek aimed to write a book on sedimentary-hosted copper deposits. Unfortunately he was unable to fulfil this desire. He passed away in Toronto (Canada) on April 20, 1991, at the age of 63.

VOCHTENITE

Renaud Vochten (°1933).

$(Fe^{2+})Fe^{3+}(UO_2)_4(PO_4)_4(OH).12\text{-}13H_2O$

Monoclinic.

Wheal Basset Mine, Redruth, Cornwall, England, U.K.

USA:
National Museum of Natural History, Washington - Type: CT: 139591 (Trelawney collection, 1978).
National Museum of Natural History, Washington - Type: HT: 362974.

IMA N°: 87-047

Zwaan, P.C., Arps, C.E.S. & De Grave, E.(1989) Vochtenite, $(Fe^{2+},Mg)Fe^{3+}[UO_2/PO_4]_4(OH).12\text{-}13H_2O$, a new uranyl phosphate mineral from Wheal Basset, Redruth, Cornwall, England. *Mineralogical Magazine,* **53**, 473-478.

http://www.handbookofmineralogy.org/pdfs/vochtenite.pdf

Aggregates of tabular reddish brown vochtenite crystals. Wheal Basset, Redruth, Cornwall, United Kingdom. R. Vochten (N° 7424-1).

Renaud Vochten was born in Deurne (Greater Antwerp) on December 8, 1933. He studied chemistry at Ghent University (GU) and graduated with a licence degree (1962). He submitted his PhD dissertation at GU (1967) and became *aggrégé* Higher Education (1976). From 1964 to 1967 he was a fellow of the Fund for Collective Fundamental Research of the Belgian National Fund for Scientific Research (NFWO-FNRS) in Brussels. During the course of 1973 and 1974, he studied mineralogy and mathematical crystallography at the University of Heidelberg, under Paul Ramdohr.

In 1967, Vochten was appointed assistant to the department of inorganic chemistry of the *Rijks Universitair Centrum* (RUCA) in Antwerp. This marked the start of a successful academic career, being appointed senior assistant in 1968, docent in 1972, professor in 1979 and full professor in 1987. He kept this position until his retirement as a professor emeritus in 1999. Between 1992 and 2000, he was a visiting professor at GU. On special request of the Ghent department of geology and soil science, he taught an optional course on gemmology and ore genesis for last-year students in geology.

Between 1962 and 1970 Vochten was involved in research on interfacial chemistry and in particular on double-layer potentials in connection with flotation problems related to ores. From 1970 onwards, his scientific work focused on uranium mineralogy and more specifically on the synthesis and structure of secondary uranium minerals. This research was carried out in collaboration with Eddy De Grave of the department of physics and astronomy at GU, Michel Deliens (*see*: deliensite), Norbert Blaton (*see*: blatonite), Paul Piret (*see*: piretite), Oswald Maurice Peeters (*see*: oswaldpeetersite), and Pieter Cornelis Zwaan of the National Museum of Geology and Mineralogy in Leiden (The Netherlands). Four of Vochten's co-authored publications describe new minerals. Their names (deliensite, piretite, blatonite and oswaldpeetersite) are an expression of Vochten's thanks and high esteem for the mineralogical research conducted by his co-workers and colleagues in the past. Consequently, it is not astonishing that in 1989 two of them (Zwaan, De Grave) proposed the name 'vochtenite' when they discovered a new iron uranyl phosphate mineral at the Wheal Basset mine, in Cornwall, England.

During his professional career, Vochten was highly interested in mineralogy. Yet, above all, he was a passionate mineral collector. Through the years, he brought together a magnificent private collection. Some time ago, he granted his collection of uranium minerals to the Museum for the History of Sciences of GU. It is presently stored at the Institute of Radiation Control in Ghent and can be visited for research purposes on request.

ACKNOWLEDGEMENTS

First and foremost, the authors wish to thank Alexander Jonckheere (Museum for the History of Sciences, Ghent University) who was in charge of the layout. Thanks to him this book and its cover are very pleasing to the eye. The authors also wish to acknowledge the aid and support of the other staff members of this museum, in particular Danny Segers (director) and Kristel Wautier (curator).

The authors are grateful to Etienne Martens for his assistance during the preparation of the crystal drawings.

Following private mineral collectors allowed to photograph some of their specimens: Blondieau Michel; Bourdeaudhui Jo; Coune Francis; Daniels Henri; Darimont Bruno; De Bondt Paul; Detaille Jean; Dillen Rik (Hendrik); du Ry Paul; Favreau Georges; Fouassin Marcel (†); Gabelica Zelimir; Gauthier Gilbert (†); Geonic Mineral Collection; Gryson Jörgen; Houssa Michel; Hubert Francis; Leemans Roger; Leydet Jean-Claude; Lheur Cédric; Lhoest Joseph (†); Mornie Jos and Nick; Pendeville Jean-Marie (†); Robert Alain; Rondelez Pierre; Schelfhout Lieve; Tossings Yves; Vandaele Jacques; Vandenberghe Luc; Vandenbosch René; van Dooren Roger; Vivier Michel.

Several Belgian mineral clubs supported the initiative of the authors by informing their members about urgent demands for information, missing specimens and documentation:
Academie voor Mineralogie (ACAM), Schoten; Association des Géologues Amateurs de Belgique (AGAB), Liège; Cercle de Minéralogie et de Paléontologie de Belgique (CMPB), Brussels; Micromonteurs de Montigny-le-Tilleul (4M), Montigny-le-Tilleul; Mineralogische Kring Antwerpen (MKA), Antwerp; Nautilus-Gent, Ghent.

Burke Ernst, former chairman of the IMA-CNMNC; De Grave Johan, Ghent University; Elburg Marlina, Ghent University; Timmermans Carla; Vercruysse Christian and Yant Philip are acknowledged for critical reading, correcting and improving specific parts of the text.

The authors would also like to thank the following institutions and universities for access to unpublished data, old documents, facilities and support: Académie royale de Belgique, Brussels; Amsab-Institute of Social History, Ghent; Brussels Royal Palace Library, Brussels; Federal Public Service Foreign Affairs, Foreign Trade and Development Cooperation: Library and Archives, Brussels; Geological Survey of Belgium, Brussels; Ghent University: Archives; Ghent University: Central Library and Library of the Department of Geology and Soil Science; Ghent University: Department of Geography, Library; Ghent University: Laboratory of Mineralogy and Petrology; Ghent University: Renard Center for Marine Geology; Göhltal Museum, Kelmis/La Calamine; Katholieke Universiteit Leuven: Archives; Katholieke Universiteit Leuven, Campus Kortrijk: Library; Koninklijke Academie van België, Brussels; Liberaal Archief vzw, Ghent; Royal Academy of Overseas Sciences, Brussels; Royal Belgian Institute of Natural Sciences, Brussels; Royal Museum for Central Africa: Department of Geology and Mineralogy, and Division of Colonial History, Tervuren; Université catholique de Louvain: Archives; University of Liège: Archives; Stadsarchief Aalst; Stadsarchief Brussels; Stadsarchief Kortrijk.

A large number of staff members and colleagues from universities, societies, institutes, museums and other institutions, and also many private persons provided the authors with useful information: Arnouts Lieve, Stadsarchief Aalst; Augustyn Ann, Stadsarchief Kortrijk; Bernhardt Heinz-Juergen, Ruhr University, Bochum, Germany; Blaton Norbert, Katholieke Universiteit Leuven; Burke Ernst, former chairman of the IMA-CNMNC, Haelen, The Netherlands; Cristofono Peter, Salem, Mass., USA; Damme Olivier, Académie royale de Belgique, Brussels; Dejonghe Léon, Geological Survey of Belgium, Brussels; Deliens Michel, Royal Belgian Institute of Natural Sciences, Brussels; Delvaux de Fenffe Damien, Royal Museum for Central Africa, Tervuren; De Paepe Christian, Katholieke Universiteit Leuven, Campus Kortrijk; de Trieu de Terdonck Eric, Angre (Honnelles); de Trieu de Terdonck Marguérite, Woluwe-Saint-Lambert (Brussels); de Trieu de Terdonck Marie-Louise, Woluwe-Saint-Lambert (Brussels); Elbers Astrid, Leiden Observatory, Leiden, The Netherlands; Ernst Gerald, Ghent University; Farges François, Museum of Natural Sciences, Paris, France; François Armand, Uccle (Brussels); Fransolet André-Mathieu, University of Liège; Goethals Herman, Royal Belgian Institute of Natural Sciences, Brussels; Groessens Eric, Geological Survey of Belgium, Brussels; Hatert Frédéric, University of Liège; Henriet Jean-Pierre, Ghent University; Hertogen Jan, Katholieke Universiteit Leuven; Jedwab Jacques, Université Libre de Bruxelles, Brussels; Laduron Dominique, Université catholique de Louvain; Lavachery Frédéric, Chaudeyrolles, France; Mees Florias, Royal Museum for Central Africa, Tervuren; Muchez Philippe, Katholieke Universiteit Leuven; Mrs. Oosterbosch Robert, Schaarbeek (Brussels); Peeters Oswald Maurice, Katholieke Universiteit Leuven; Perroud Pierre, Athena Mineralogy, Switzerland; Mrs. Piret-Meunier Paul, Nil-Saint-Vincent-Saint Martin (Walhain); Postl Walter, University of Graz, Austria; Rämö Tapani, University of Helsinki, Finland; Segers Vera, Ghent University; Streel Maurice, University of Liège; Tack Luc, Royal Museum for Central Africa, Tervuren; Uher Pavel, University of Bratislava, Slovakia; Van den haute Peter, Ghent University; Van Tassel René, Royal Belgian Institute of Natural Sciences; Van Wambeke Léopold, Kraainem; Mrs. Verbeek Theodoor, Toronto, Canada; Verreecken Emiel, Dienst Ruimtelijke Ordening van de stad Aalst; Vochten Renaud, Rijks Universitair Centrum Antwerp; Vurro Filippo, Università degli Studi di Bari, Italy; Warin Roger, University of Liège; Zelenski Michael, Institute of Experimental Mineralogy, Chernogolovka, Russian Federation.

Both the authors and the editors would like to express their gratitude to the publisher, Academia Press, and Ghent University for their support during the realization of this work and their generous investment in this project. The editors would also like to thank "Wetenschap maakt Knap" and "Vlaanderen in Actie" for their financial support to the project.

PORTRAIT CREDITS[1]

Bia, Lucien: © *Union Minière du Haut-Katanga*, Brussels
Billiet, Valère: © Archives of Ghent University, Ghent
Blaton*, Norbert
Borgniez, Georges: © *Faculté Polytechnique de Mons*, Mons
Buttgenbach, Henri: © *Académie royale de Belgique*, Brussels
Camerman, Carl: © *Société Belge de Géologie, de Paléontologie et d'Hydrologie*, Brussels
Cattier, Félicien: © Royal Belgian Colonial Institute, Brussels
Cesàro, Giuseppe: © *Académie royale de Belgique*, Brussels
Comblain, Gordon: © Archives of the Royal Museum for Central Africa, Tervuren
Cornet, Jules: © *Académie royale de Belgique,* Bruxelles. Engraving by Montenet
Cousin, Jules: © Royal Academy of Overseas Sciences, Brussels
de Koninck, Laurent-Guillaume: © *Académie royale de Belgique*, Brussels. Engraving by Gustave Biot
Delhaye, Fernand: © Royal Belgian Colonial Institute, Brussels
Deliens*, Michel
Delvaux de Fenffe, Jean Charles: © Delvaux de Fenffe Damien, Royal Museum for Central Africa, Tervuren. Litho by Schubert, after a painting by Barthélemy Vieillevoye
Derriks, Jean-Marie: © *Association des Géologues Amateurs de Belgique*, Liège (courtesy of Roger Warin)
De Boodt, Anselmus Boetius: Engraving by Egidius Sadeler, ca 1598
de Trieu de Terdonck, Robert: Courtesy of Eric de Trieu de Terdonck, Angre (Honnelles)
Dewalque, Gustave: © *Académie royale de Belgique*, Brussels. Engraving by Louise Dance
De Windt, Jan: © *Ligue du souvenir congolais*, Brussels
Diderrich, Norbert: © *Le Mouvement géographique, Journal populaire des Sciences géographiques*, Brussels
d'Omalius d'Halloy, Jean-Baptiste: © *Académie royale de Belgique*, Brussels. Engraving by F. de Meersman
Donnay, Joseph: © The American Mineralogist, Washington, DC, USA
Droogmans, Hubert: © *Union Minière du Haut-Katanga*, Brussels
Drugman, Julien: © *Société française de Minéralogie et de Crystallographie,* Paris, France
Dumont, André: © *Académie royale de Belgique*, Brussels
Fourmarier, Paul: © Royal Academy of Overseas Sciences, Brussels
Fraipont, Julien: © *Académie royale des Sciences, des Lettres et des Beaux-Arts de Belgique*, Brussels
François*, Armand
Fransolet*, André-Mathieu
Graulich, Jean-Marie: © *Geologica Belgica*, Brussels
Jedwab*, Jacques
Lepersonne, Jacques: © Archives of the Royal Museum for Central Africa, Tervuren
Lohest, Maximilien (Max): © *Académie royale de Belgique*, Brussels. After a litho by J. Gers
Marthoz, Aimé: © Royal Academy of Overseas Sciences, Brussels
Mélon, Joseph: © *Geologica Belgica*, Brussels
Meyer, André: © Archives of the Royal Museum for Central Africa, Tervuren
Moreau, Jules © Archives of the *Université catholique de Louvain,* Louvain-la-Neuve
Mund, Walter: © *Académie royale de Belgique*, Brussels
Oosterbosch, Robert: Courtesy of Mrs. Robert Oosterbosch, Schaarbeek (Brussels)
Peeters*, Maurice Oswald
Piret, Paul: Courtesy of Mrs. Paul Piret-Meunier, Nil-Saint-Vincent-Saint-Martin (Walhain)
Renard, Alphonse: In: Greindl, L (1907): *Le mouvement scientifique en Belgique. Les sciences géologiques.* p.34, Bruxelles: Oscar Schepens & Cie
Renier, Armand: © *Académie royale de Belgique*, Brussels
Salée, Achille: © Archives of the *Université catholique de Louvain,* Louvain-la-Neuve
Schoep, Alfred: © University Library, Ghent, Collection of Portraits
Sengier, Edgard: © *Union Minière du Haut-Katanga*, Brussels
Stainier, Xavier: © Archives of Ghent University, Ghent
Stas, Jean-Servais: © A.C.L. Brussels. Painting by Louis Boschmans, 1892, Leuven, Museum Vander Kelen-Mertens (M Museum)

(1) *Surnames followed by an asterisk in superscript refer to living persons who supplied the authors with a photograph.*

Tazieff, Haroun: © Courtesy of Frédéric Lavachery, Chaudeyrolles, France
Thoreau, Jacques: © Archives of the *Académie royale des Sciences, des Lettres et des Beaux-Arts de Belgique*, Brussels (n° 15465)
Vandendriessche, Adriaan: © *Natuurwetenschappelijk Tijdschrift*, Ghent
Van Meerssche, Maurice: © *Académie royale de Belgique*, Brussels.
Van Tassel*, René
Varlamoff, Nicolas: © Royal Academy of Overseas Sciences, Brussels
Verbeek, Theodoor: Courtesy of Mrs. Theodoor Verbeek, Toronto (Canada)
Viaene, Willy: © *Katholieke Universiteit Leuven, Aardkundige Mededelingen*, Leuven
Vochten*, Renaud

PERSONAL NAMES INDEX[1]

(1) Names highlighted in **bold** *type refer to individuals with a biographical description. The page number depicting the portrait of the individual (when available) is underlined.*

MINERAL NAMES INDEX[1]

[1] *Minerals with Belgian roots are highlighted in* **bold** *type (IMA approved species), in* ***bold italic*** *type (questionable species), or in regular italic type (discredited species). Page numbers which are underlined show at least one photograph of the referred mineral. Minerals without Belgian roots (IMA approved or not) are printed in* regular *font style.*

COVER CAPTIONS

Painting of the zinc ore exploitation at Altenberg/Vieille-Montagne, Kelmis/La Calamine, by Jean-Baptiste Bastiné (1843).
Courtesy of Göhltal Museum, Kelmis/La Calamine, Liège Province, Belgium.

Hopeite crystals, collection of the Royal Belgian Institute of Natural Sciences, Brussels.

SEM photograph of tazieffite.
Courtesy of Michael Zelenski, Chernogolovka, Russian Federation.

Mutnovsky volcano, type locality of tazieffite, Kamchatka Peninsula, Russian Federation.
Courtesy of Michael Zelenski, Chernogolovka, Russian Federation.

SOCIÉTÉ

Mines et Fo

LA VIEILLE

Dixième

(L'Action A entière a une vale

TITRE N° 170,0

Deux Administrateurs,

Liège, le 2

H. CATENACCI. DEL.

Paris. — Imp. Paul